INTERPLANETARY ROBOTS

ALSO BY ROD PYLE

Amazing Stories of the Space Age

Curiosity

Destination Mars

INTERPLANETARY ROBOTS

TRUE STORIES OF SPACE EXPLORATION

ROD PYLE

59 John Glenn Drive
Amherst, New York 14228

Published 2019 by Prometheus Books

Inquiries should be addressed to
Prometheus Books
59 John Glenn Drive
Amherst, New York 14228
VOICE: 716–691–0133 • FAX: 716–691–0137
WWW.PROMETHEUSBOOKS.COM

23 22 21 20 19 5 4 3 2 1

Library of Congress Cataloging-in-Publication Data

Names: Pyle, Rod, author.
Title: Interplanetary robots : true stories of space exploration / by Rod Pyle.
Description: Amherst, New York : Prometheus Books, 2019. | Includes bibliographical references and index.
Identifiers: LCCN 2018038827 (print) | LCCN 2018040090 (ebook) | ISBN 9781633885035 (ebook) | ISBN 9781633885028 (paperback)
Subjects: LCSH: Astronautics—United States—History. | Space robotics—United States—History. | United States. National Aeronautics and Space Administration.
Classification: LCC TL789.8.U6 (ebook) | LCC TL789.8.U6 P95 2019 (print) | DDC 629.43/54—dc23
LC record available at https://lccn.loc.gov/2018038827

Printed in the United States of America

To the memory of Ron Ding,
a man of brilliance and humility,
who saw the future in all its beauty.
You left us too soon for your own journey into the great beyond.

CONTENTS

FOREWORD

The historic Soviet Sputnik mission in 1957 began a spectacular era of space exploration. We found out right away, with many early failures, that space exploration is a hazardous adventure, and even today we aren't able to perfectly accomplish each mission. Before Sputnik, everything we knew about our solar system came from ground-based telescope observations and from analysis of meteorites. The space program changed that forever, since it provided us the opportunity to study our solar system more thoroughly, to travel to distant objects within it and to study those objects up close to discover how the solar system was created and how it has evolved. With the ever-increasing ability to reach and navigate to more distant objects, we began a series of missions with always new and better instruments. Each new mission built on what we had learned from its predecessor. In order to thoroughly explore an object of interest, we developed a mission sequence—flyby, orbit, land, rove, and return samples—with each step creating new technologies and tackling new challenges in order to accomplish the next mission. At times we would be so bold as to skip some steps, like landing the Huygens probe on Titan without having the benefit of a precursor orbiter providing us with detailed surface maps and allowing the selection of a safe landing site beforehand.

What a historic time we have just completed and yet, in many ways, it is just the beginning. With the tremendously successful flyby of the Pluto system by the New Horizons spacecraft in July 2015, humankind completed its initial survey of the solar system within sixty years. Solar system exploration has always been, and continues to be, a grand human adventure that seeks to discover the nature and origin of our celestial neighborhood and to explore whether life exists or could have existed beyond Earth. Rod's book captures this excitement, reliving missions that NASA and the Soviets launched, and some that never left the drawing board, while also looking into the immediate future for our next steps. I, too, am fascinated by space history. We all know that history will repeat itself if we let it, and we just have to avoid our past mistakes. As Rod so aptly put it, this is "a magnificent tale of exploration and discovery . . ." It is fully worthy to be examined and enjoyed.

James Green
NASA Chief Scientist

INTRODUCTION

"You maniacs! You blew it up! Oh, damn you . . . God damn you all to hell!" and so on, rants Charlton Heston at the end of *The Planet of the Apes*. That's at the end of the 1968 classic film, in the scene where he's discovered that he is stranded not on some strange planet enlivened by parallel evolution, but has in fact returned home to a post-nuclear-holocaust Earth.

I thought much the same, however, on July 15, 1965, when the first images of Mars were returned to Earth.

This book got its start on that day, when Mariner 4 destroyed the Mars of the Victorians and the romantics. In mid-July of that year, the first American probe to reach the planet flew by, sending back twenty-two black-and-white images over the next few days. While grainy and low-resolution, those few images—showing less than one percent of the surface—changed everything. That was the day that the Mars of Giovanni Schiaparelli, Percival Lowell, H. G. Wells, and Edgar Rice Burroughs, to name just a few of the many who had dreamed about Mars for decades, vanished. In place of those dreams was a dry, cratered, and frigid desert of a planet. The atmosphere was much thinner than we thought, the temperatures lower than we thought, and the presence of standing liquid water impossible. Bye-bye, friendly neighbors on another world. Farewell vast oceans and sprawling forests. See you later, little green men. Ta-ta, advanced alien civilization and your planet-girdling canals.

I was just eight years old at the time—too young to fully understand the more complex ramifications of those first images of Mars, but surely old enough to know that the red world imprinted on my brain by the likes of Ray Bradbury and Wernher von Braun—visionaries at the extremes of our Martian visions—had vanished forever.

What did we get instead of Martian canals, neighbors in space, and a second home? We got the truth—at least a part of it. Over the next few decades, in remarkably short order, more planetary explorers—robots sent mostly by NASA, but also by the Soviet Union—reconnoitered the red planet and slowly revealed more of its secrets. Mars is indeed dry, desolate, and windswept, but it holds far more water than could have been expected, locked up in the poles and in subsurface glaciers, and it has a much more complex geological history than was first hinted at by Mariner 4. That first look took something away from us, but the many orbiters and rovers to visit the planet since have given much back. Mars is a compelling, if forbidding, place.

And so it has been with the rest of the solar system. It is at the same time far less friendly and hospitable than we had hoped, but more fascinating than we could have imagined. It took over fifty years to reach the rest of the planets. Pluto (once a planet, now classified as a dwarf planet) was imaged by the New Horizons spacecraft in 2015, using a flyby trajectory hauntingly like Mariner 4's. That was the last of them—every planet from Mercury, nestled next to the sun, to Neptune and even (dwarf planet) Pluto had been imaged and investigated. We had done the easy part. Now it is time to go back to the most compelling places we've seen—the moons of Jupiter and Saturn, with their warm, subsurface oceans; the subterranean glaciers of Mars; the hydrocarbon oceans of Titan—and explore these places in detail. Answers about the very origins of life in our solar system may await us there, and the most primary questions concerning our existence and place in the universe may be answered.

This book is the story of the exploration of the solar system, told in historical detail with occasional "flash forwards" to missions planned for the near future. Like its companion book about the human spaceflight programs of the last sixty years, *Amazing Stories of the Space Age*, the book is a mixture of programs that were flown in the past, some that were planned but never flown, and others that will be flown in the near future. *Interplanetary Robots* covers primarily the efforts of the US and Soviet space programs, as those two superpowers have done most of the heavy lifting in planetary exploration to date. It has been impossible to write about every mission and program—there are simply too many to document in a single book—but the major ones are discussed.

Interplanetary Robots has been a guilty pleasure to research and write, but as with all such efforts there will be errors within. I take responsibility for most of them. It's traditional for authors to say "any errors are mine," but, in truth, go to any five primary sources while researching the space age—go ahead, I dare you—and you will invariably find disagreement on dates, spacecraft mass, and a lot of variability in who-said-what and when. Add to this the occasionally inaccurate, fading memories of people who were there, working on the programs, who have been interviewed, and you can understand how errors will sometimes creep in. It is inevitable, and all that space authors can do is consult multiple sources, ask knowledgeable friends to proofread the book, and hope. After all, the men and women who were building and flying these amazing machines were not, for the most part, thinking about perfect recordkeeping for the historians who would come later—they were thinking about getting the job done, flying the mission, and gaining the maximum science return from the heavy investment of tax dollars. Records can be incomplete at times, and some disagree with each other, but that's not really the point.

At the end of the day, space exploration is not about dates and times, weights and measures, or facts and figures. It is a magnificent tale of exploration and discovery, and of the many thousands of incredible people who are behind these endeavors. It is about a compelling story well told, and that is what I've attempted to do. Other than that, any errors are . . . all mine (except for the ones all those other people made).

Please read and enjoy. If you like what you see—or in the unlikely event that you don't—feel free to reach out and let me know on Facebook or via my website, www.pylebooks.com. And don't forget to leave a few reviews at Amazon, Barnes and Noble, and the other online book outlets—authors live and die by reviews. I do read them and respond in time.

And now, let's take a trip through the solar system with the people and machines who have explored it with ingenuity, drive, and passion, as we join with the interplanetary robots on their incredible journeys into the great void beyond earth orbit. It's a voyage well worth taking, and I assure you that your time spent with the remarkable people and machines who have journeyed to its vast reaches will be well spent.

Chapter 1

ALONE IN THE DARK

If you were to somehow find yourself floating in deep space sometime in 2024, about ten billion miles past the orbit of Pluto, and you were in the perfect spot, you would be in for an extremely brief treat. Swiveling your gaze away from the starry panorama of the Milky Way and beyond, looking back from where you came, you would see a dim star—somewhat brighter than the others, but unremarkable, and that would be our sun. Earth would not be visible without a powerful telescope. But wait one moment . . . there is a tiny, dark speck coming toward you, almost blotting out the wan image of the sun.

In the time it takes to write *this*, a timeworn spacecraft hurtles past you, traveling at some 38,610 miles per hour—and then it's gone into the black void. That was Voyager 1, which has been speeding through space since 1977. The probe passed the distance of Pluto, about 4.67 billion miles from the sun, back when big hair bands like Van Halen still topped the music charts in the 1980s. It traversed the vaguely defined edge of the solar system in 2012, and continues its journey past you into interstellar space, the void between stars.

Why did I pick 2024 to thrust us into this chilly spot in the void beyond the solar system to witness this blistering flyby? Because sometime that year, or perhaps the next, or the one after that, will be the end of the history-making missions of Voyagers 1 and 2. At that time, the available power from its plutonium-238 fuel source will be too weak to power its instruments. The dwindling radio signal from the spacecraft will also be increasingly difficult to track from Earth, now over 15,500,000,000 miles distant. That nuclear power supply will continue to generate heat, which it converts to electricity, for hundreds of years, but its ability to create useful levels of electrical current will be long gone by then. While the plug of plutonium will still be hot for the rest of the twenty-first century, the devices that convert that heat, called thermocouples, will have long since degraded in their ability to do their job; and the current produced while they are still active will be too little to run the spacecraft. Hence the mission cutoff date of sometime in the mid-2020s.

Fig. 1.1. "Here comes Voyag—I mean, there goes Voyager." At 38,000 miles per hour, you would not have long to snap this photo as it streaked by. Image from NASA.

As of 2019, the Voyager 1 and 2 continue their quiet journey past the rim of the solar system, venturing through the great emptiness beyond. Their final encounter with a planet was that of Voyager 2, which screamed past Neptune in August of 1989. The probes had to work quickly during these planetary flybys—it was only later, with spacecraft like the Jupiter-bound Galileo and Cassini, which flew past Saturn, that these giant machines were actually able to orbit the massive worlds they studied. But before that went Pioneer 10 and 11, the aptly named first spacecraft to traverse the outer planets, and the Voyagers were the first to do detailed work there. Much later, Voyager 1 was the first to pass beyond the edge of the solar system, also known as the heliosphere. (Voyager 2 had a longer route to follow, and is just crossing that thick boundary layer now.) The Pioneers and the Voyagers will continue their journey through space for billions of years, passing distant star systems in the future: Voyager 1 will pass a star named Gliese 445 in about 40,000 years, at a distance of about one and a half light years, and Voyager 2 will pass another star called Ross 248 by nearly the same distance at about the same time.

But both started out from the Earth, and are the children of NASA's Jet Propulsion Laboratory (JPL), which is where our story begins.

Chapter 2

THE CENTER OF THE UNIVERSE: PART 1

August 5, 2012, is yet another hot August day in Southern California. While the entire region sits in a smoggy haze throughout the summer, the San Gabriel Valley, Los Angeles's northeastern quadrant, generally gets the worst of it. The prevailing winds often blow in from the ocean, about twenty miles distant as the crow flies, moving past the complex of freeways that crisscross the megalopolis. This breeze scrapes up the emissions from the millions of cars that sit in twenty-four-hour gridlock across the city. That smog warms and darkens through photochemical reactions with the sun and is slowly pushed over the mid-city hills to settle into the inland valleys. Once there, it heats, and sits, and heats some more, becoming increasingly acrid with each passing afternoon hour. This day is no exception, but at least it is not roastingly hot. While the summer heat here can easily top three digits by egg-frying margins, the temperature on the fifth is only about 82 degrees, with low humidity—in downtown Los Angeles anyhow. Cross into the valley and the temperatures magically elevate by an average of ten degrees. Hot indeed, but not too daunting for a quick drive up to NASA's Jet Propulsion Laboratory, just a few miles northwest from my home, on this auspicious day.

As I head up to "The Lab," as it's often referred to in these parts, the sky overhead is *almost* blue—not a small accomplishment for midsummer, when LA smog peaks. JPL is tucked into the southern slopes of the San Gabriel Mountains, and for the past four months the scrub brush surrounding it has been brown and dead—Southern California is in yet another period of prolonged drought. A few scruffy trees struggle to cling to life in the arid climate, creating a vague green halo around the edge of the JPL campus.

Fig. 2.1. NASA's Jet Propulsion Laboratory is nestled in the San Gabriel Mountains, not far from Pasadena, California. This is a rare smog-free photo. Image from NASA.

Getting off the freeway, the traffic lightens and the midday trek up the steep street that dead-ends at JPL's front gate is a quick drive. While the NASA center's mailing address is in Pasadena, California, its physical location lies in a suburb called La Cañada Flintridge. These sprawling postwar communities are chock-a-block with small two- and three-bedroom bungalows built to accommodate returning Second World War GIs and their rapidly expanding families in the 1940s and 1950s, and were an affordable alternative to the more expensive regions closer to metropolitan Los Angeles and their tony Western relatives of Hancock Park, Beverly Hills, and Santa Monica. But this is no longer the case—the *average* price of a plain-Jane ranch-style home of about 1,600 square feet in this locale in 2012 soared to nearly a million dollars (and in 2018 it's almost $2 million). It's a high price to pay to live in the overheated northeastern suburbs, but with JPL resting in a crook in the northern foothills, the only areas that don't require a lengthy commute are here and nearby Pasadena or La Crescenta. During rush hour, a trek from the more affordable communities to the east, let's say up to about thirty miles away by freeway, can take well over two hours one way.

The final leg of my journey is a steep climb up Oak Grove Drive, which earns its name from the procession of old oak trees that line its sides. The cooling shade they provide is a welcome relief for the (mostly younger) engineers and scientists that troop up the hill by foot every day—the workforce tends to take environmentalism seriously, as many of their Earth-orbiting machines report back the bad news on climate change daily. They are a committed bunch, their well-worn Nikes and Birkenstocks a testament to their commitment to environmentalism.

As the road ends at the boundary between the city and federal NASA property, there is a Jetsons-style guard post. It looks like a space age submarine conning tower with windows, with a huge aluminum airplane wing suspended above. A US flag hangs forlornly in the sullen summer air, and a bright blue NASA "meatball" logo is affixed to the top of the structure. Emblazoned below that is "Welcome to Our Universe." It's a nice sentiment that belies how difficult it is to get into this place—if you are not an employee or invited visitor, access is rare and can take months. Notably, JPL is one of the only NASA field centers in the country that does not have a museum attached and open to the public—they do have a small historical center inside, with spacecraft mockups beautifully displayed, but entrance is by official clearance only. There is also an annual two-day open house—the most recent had been just weeks before—but the 18,000 free tickets are snapped up online in less than an hour. It is one of the most popular event of its kind in the country. In short, my visit is a privilege.

I'm here on a press pass to witness the landing of the Mars Science Laboratory, more commonly known simply as the Curiosity rover. The landing itself will be made famous by the national media, and by a JPL-produced short video called "Seven Minutes of Terror." The title refers to the Entry-Descent-and-Landing (EDL) phase of Curiosity's mission, during which the rover will enter the Martian atmosphere and, hopefully, land safely on the surface of the red planet. The video will go viral almost instantly, earning over millions of views within weeks. It masterfully and dramatically conveys the dangers of attempting to land a large, heavy machine on Mars with existing technology, and it outlines the stakes of the mission vividly. As of this afternoon, however, nobody knows what will happen—the big event is still hours away. But we are all well aware of the challenges.

Once past the guard post, I am directed to visitor parking. The main visitor lot is full of news vans, their satellite dishes poking high into the arid skies like sunflowers seeking a bit more light. The reporters have gotten here early in an attempt to capture some of the "drama" of the event. I wish them luck—this is not a community given to stray outbursts of emotion, at least not until the final touchdown. The TV technicians lounge around the vans, making sure the satellite links are working, and the reporters are scattered around the main courtyard within JPL. Their access will be limited to the visitor's center and the open-air main quad—anywhere else requires higher clearance than a press pass and visitor badge.

I follow the green painted line over the baked-out asphalt to the "Western Annex." It's the parking lot furthest from the main entrance, and even that is nearly full. The parking is terraced up the foothills for perhaps a quarter mile, and I know there is more on the eastern perimeter, but this is where I've been directed. I squeeze my bulky SUV in between two sparkling imports that are worth easily ten times what my Tahoe is . . . they have employee tags hanging from their rearview mirrors. A job here looks better and better.

I'm on assignment for Space.com, a premiere outlet for space news online. They are a great bunch, and while the work does not pay handsomely it is useful for gaining access to places that would be otherwise off-limits. I'll be shooting video tonight, so I grab my heavy rig—this is circa 2010 equipment, and many times heavier than what is used by most videographers today. Hoisting my camera bag and beefy tripod (which, come to think of it, is also worth more than my car), I trudge the third-of-a-mile walk to the visitor's center where I will check in. It seems farther in the heat. As it turns out, I'll become familiar with this walk in a few years—once employed here, one discovers that successful parking is the realm of the early riser, and book authors who often toil well into the dark hours and tend to show up for work after 10:00 a.m. are forced to park far out on the perimeter of the facility.

After clearing visitor control, I step into the mall, the main open area on the JPL campus. The word "campus" is well used here, because it tends to bring academia to mind. If you went to college, there was probably a large tree-shaded area somewhere. JPL has the same thing—it's called "the mall" here—and it's no accident. The Lab is unique among NASA's ten field centers located across the United States in that it is not run directly by NASA. JPL is *funded* by NASA, but that funding goes to the California institute of Technology, or Caltech, in Pasadena, which manages JPL. This university connection is a contrivance of history, and has roots in JPL's origins, but the net effect is that JPL workers are not civil servants like other NASA employees, they are Caltech employees. A substantial portion of JPL's workforce consists of outside and independent contractors as well, who do not share in Caltech's full-time employee benefits, and this saves NASA money. Another effect of this unique relationship is that JPL looks and feels a lot more like a college campus than the other NASA centers, now showing their age in an increasingly inelegant fashion. JPL may be rough around the edges, but no more so than many public universities.

At the center of the mall is a fountain, another apparent relic of the 1960s, that looks a bit like a giant cement kidney with a hollow center and water squirting skyward from a few jets. It's nice because it makes calming sounds and hydrates the dry LA air, but it won't be doing so for long. A few years after Curiosity's landing, the fountain was shut off due to water conservation efforts. Apparently, even hyper-chlorinated water that is recycled through an endless loop in the cement kidney is too much to spare during LA's permanent drought. Most were sad to see it go—today it is a dry concrete trough. But NASA and Caltech are both very sensitive to public sentiment, and water conservation is all the rage in gulch-dry LA, so *c'est la vie.*

Nearby are large, grassy areas with chairs and lovely wood tables with umbrellas—intimate seating areas that offer little pools of shade where people can gather to discuss their latest plans for the conquest of space, or just grab some private time. There's a Starbucks stand near the fountain, and a couple of cafeterias on the north side of the campus—altogether very civilized and college-like.

The rest of JPL continues east and north from this area, with some of the newer buildings along those axes. It's not what one would call pretty—this place is all about utility—but at least some of the more recent structures have more windows than the older buildings. The number of offices in this place that have no windows, not even in the metal doors enclosing the offices within, is stunning. The desks are gray metal government issue, the walls covered in burlap or light gray paint (or perhaps it's taupe . . . who can tell?), and the floors are covered in well-worn, close-pile industrial-grade indoor-outdoor carpet or linoleum. Much of JPL could be the interior of the Pentagon or an elderly aerospace contractor, but there are exceptions.

One exception is the main administration building, which has the uninspiring name of Building 180. This is where the leaders of robotic space exploration ply their trade, and do so, for the most part, with impressive efficiency and vision. The place is generally run as a tight ship.

The upstairs offices in 180 tend to have sweeping views of either the valley to the south, or the mountains to the north. Both views look less withered and arthritic from up here—maybe it's the sun-shading Mylar coating on the windows. With the AC on you can almost imagine nice weather out there. Almost.

The lobby of Building 180 contains two remarkable things. First is a clone of the Curiosity rover, which will be landing later on this day in 2012, sitting against the east wall with a couple of soft spotlights on it. It is magnificent, due partly to its size—that of a small car—its complexity, and the audaciousness of its engineering. The second amazing thing is an art installation against the north wall. This is the handiwork of Dan Goods, a "Visual Strategist" at JPL. What is a visual strategist, you might ask? He is an artist with some engineering smarts, a graduate of the nearby Art Center College of Design, who specializes in communicating the incredible accomplishments of robotic space exploration in mostly nontraditional ways. He heads a small design studio housed in a collection of rather ramshackle "temporary structures"—think interconnected mobile homes—up the hillside from here. On the outside they look like short-term military housing, but inside they are painted flat-black and contain magic. This installation is a bit of that magic.

Fig. 2.2. Curiosity's unflown twin inside Building 180 at JPL. The top deck is at chest height, so it's bigger than it looks. Image by Rod Pyle.

Goods has created a light sculpture in Building 180 that, in one small space, captures the wonder of planetary exploration. It is called, simply, "The Pulse," and runs from the floor to the lobby ceiling, a twenty-foot pillar of LED light strips in a cylindrical shape, about five feet across. It doesn't look like much at first—arrive before it starts up, and you stand there, wondering what this set of gray strips running floor-to-ceiling next to the Mars rover are. Then it suddenly bursts into cold white light, a massive wave of it, moving from top to bottom or the reverse. The pulses of light continue, then reverse, then repeat. Eventually, tilting your head a bit, you make out words spelled in these fleeting bursts—"VOYAGER" or "ODYSSEY"—and realize that what you are looking at is a visual representation of communications between JPL and its many robotic explorers throughout the solar system and, thanks to Voyager, beyond. Out go the digital bits of orders from the engineers and scientists, and back come the responses—acknowledged commands or gigabytes of science data. While the receiving stations are far away, scattered across the globe, this simple art-by-light installation elegantly communicates the breadth of the US planetary exploration program in one place. It's great fun to hang around the lobby and watch new visitors as they try to figure out what they are looking at. It can take a while, but there is always that special moment when a smile crosses their faces—it's a distinctly emotional response to technology that Goods and his small team are singularly talented at engendering in both laypeople and hardened technocrats alike. There should be one of these in every shopping mall and library across America, I muse on that day. It could be NASA's best calling card to the public.

Next to this artwork is a simple plaque on the wall that says, "Dare Mighty Things." The words are not NASA's; rather, it is a quote from Theodore Roosevelt from this passage: "Far better is it to dare mighty things, to win glorious triumphs, even though checkered by failure . . . than to rank with those poor spirits who neither enjoy nor suffer much, because they live in a gray twilight that knows not victory nor defeat."[1] While the sentiment may imply more of a martial spirit than intended by the space agency, the words are inspiring.

Moving back to the mall, what sets today apart from normal operations at JPL is apparent: the open concrete plaza just inside the main entrance is festooned with awnings, fluttering banners, banquet tables, informational displays, and spacecraft mockups. The Curiosity rover will be reaching Mars tonight, and will either land magnificently, or crash spectacularly, before this day is out. The media is here in droves for the victory, or the inquest, and must be engaged and entertained until the big show later this evening.

It's a well-designed display of media savvy on NASA's part. JPL has always been good at that, better than most of NASA by some estimations. Possibly it is the proximity to Hollywood, or maybe due simply to their smart and creative communications staff, but they somehow make it look easy. It's not, especially given very constrained budgets—the robotic planetary programs created here don't get a lot of NASA's money. Robotic programs dominate JPL, and include downward-looking Earth science missions in addition to their work at Mars, the outer solar system, and the moon and sun. JPL's annual budget is about $1.6 billion, a bit less than one-tenth of NASA's overall spending.[2] So perhaps this has resulted in a bit of scrappy can-do-ism on the part of Caltech's jewel-in-the-crown, but that's all conjecture. Suffice it to say that they are putting on a very good show today.

The centerpiece of the displays is the full-sized Curiosity mockup. Off-duty engineers and technicians take turns giving talks and answering questions. The media crowd is getting thicker as the afternoon wears on. The landing of the robotic rover is scheduled for about 10:30 this evening, but the news cycle will pick up on the story earlier, as Mars looms increasingly large in the simulated graphics that decorate the many video screens stationed around the lab. This display makes the hurtling Curiosity look like a gray metallic bug about to splatter against a giant ruddy windshield. Small clumps of reporters look on with varying levels of comprehension—the magazine and newspaper writers have a more intense, concentrated expression than most of the television reporters, who chat impatiently with their videographers.

As the sun drops, the mall becomes increasingly pleasant. It's a lovely place to spend a summer evening, and the crowd of both JPLers and media people divide their time between here and Von Karman

Auditorium, where the press area is set up. The real action is occurring a few hundred yards to the north, in the building inelegantly named the SFOF, or Space Flight Operations Facility (yes, NASA could benefit from consultation with some expensive branding experts regarding names), but Von Karman is about as close to mission control as the bulk of the media is going to get tonight.

Of course, given my choice, I'd go up to the SFOF. It's a multistory structure just east of Building 180, and is the hub of JPL's flight activity. The SFOF is the robotic spaceflight equivalent of Houston's Mission Control, from which the human spaceflight program has been operated since the days of the Mercury program in the 1960s. JPL's facility is a bit smaller, a bit less impressed with itself (Houston's "Mission Control" is capitalized by decree of NASA's media style guidebook; Pasadena's "mission control" is not). But it is much cooler to look at than Houston's: all blue-lit glass and blue-gray carpets, with dozens of multiscreen control areas throughout. Above it all are big-screen projections. To the left and right are screens with mission clocks; another has multiple icons for radio receiving dishes all over the world and shows a graphic indication of which dish is receiving or transmitting to which spacecraft. Another shows the CGI view of Curiosity closing on Mars, and yet another simply has raw data. It's a visual treat; an overwhelming collection of sights and colors worthy of Stanley Kubrick (or for a younger generation, J. J. Abrams, without the lens flares).

At the approximate center of the control room, between the four or five long rows of control stations, sits a plaque mounted under glass in the floor. Here is where JPL's humor—and perhaps a bit of hubris—shines through. It says, simply, "The Center of the Universe," and announces that this is the exact spot at JPL where planetary exploration happens. The Johnson Space Center, or Kennedy Space Center, or NASA Headquarters might disagree, but I'm happy to let JPL be the universal hub. It gives Pasadenans something to obsess about beyond the Rose Parade and the rowdy football game that follows it on each New Year's Day.

Back over in Von Karman, we of the media are doing our best to stay out of each other's way, without a lot of success. If you cover these kinds of events very often, you build up a little private shell—a zone of exclusion, if you will—from which you can ignore most of what's going on around you except for the story you are there to cover. I was startlingly reminded of this a few years later when reporting on the unveiling of the new and improved Dragon 2 space capsule over at SpaceX, Elon Musk's dynamic new rocket company just across town. At that event, representatives of the media were jammed onto a tiny raised stage behind the larger, more relaxed VIP area down front, all inside SpaceX's sprawling rocket factory. The "media box" was all elbows and tripods, each of us vying for a tiny patch of the dais from which to video the Great Unveiling. It was pure Musk, with purple lights and fog machines and mirror balls, as the draping fabric was yanked from his latest, almost impossibly cool creation; but I barely got the shot thanks to the clambering press. At least here, at Von Karman, you have some breathing room.

As we await the "Seven Minutes of Terror" landing sequence, there are a few places to wander and wonder. Next to the main auditorium is an ad-hoc press room, which has a few dozen temporary desks where reporters can plop down their laptops and file stories. We're shoulder to shoulder over there, but it's a place to write, and damn if they don't have free Wi-Fi. The room is normally part of their small museum, with a full-sized mockup of the Galileo Jupiter probe dominating the center, and a Mars Exploration Rover, Curiosity's smaller predecessor, up front. The walls all around are covered by backlit graphics of various JPL undertakings, and at the center, near the Galileo probe, are large video screens with the now-familiar face of Mars looming large from Curiosity's presumed perspective.

Curiosity is being drawn in by Mars's gravitational field, and the moment it encounters the thin Martian atmosphere it will be traveling at just under 13,000 miles per hour. That's really fast, and leaves no room for error. The first Mars landers, the Vikings of the 1970s, were flown from Earth and into orbit around Mars, to loiter there for a month until flight planners had a chance finalize their selection of a landing spot. Not so Curiosity, or any of the Mars landers since Viking. Carrying the extra rockets to slow a spacecraft into Martian orbit after a roughly 350,000,000-mile journey means launching a lot of extra weight (the straight-line distance varies from about 33 million miles to 254 million, but due to the long, curving route that the probes must follow from the Earth to Mars, the flight path is longer). Add to that the fuel required to fire those rockets, and it results in a much larger spacecraft, and one much more expensive to design, build, and launch. Since Mars Pathfinder, the first rover to reach Mars in 1997, and the first landing there since Viking 2 in 1976, all Mars landers and rovers have used the "hunting rifle" approach: they are shot from Earth by a rocket, travel a long arcing route to the red planet, and hurtle straight into the atmosphere at high velocity—no errors permitted. The angle of entry must be absolutely correct, and all the steps required to land the spacecraft executed perfectly, for success. This is one reason that, for the first few decades of Martian exploration, almost half the attempts failed—everything must be perfect. In fact, until the twenty-first century, only the United States had succeeded at putting a fully functioning spacecraft into Martian orbit (the Soviet Union, and later Russia, has tried dozens of times with only limited success), and even now, the United States is the only spacefaring power to successfully land a spacecraft onto the surface of Mars that has operated for more than a few seconds.

As they like to say at JPL, Mars is hard.

The hours are passing slowly, and Mars is getting awfully big on that screen. The moment of truth will soon be upon us.

Chapter 3

EARLY PLANS

At the dawn of the space race, and early in the Cold War, the militaries of the Soviet Union and the United States were continually sizing each other up, using whatever intelligence-gathering methods they had at their disposal—espionage, high-altitude overflight, honeypot seductions, and whatever other methods would yield some clue as to what the enemy might be up to.

This was just a few months after the Russians had embarrassed the United States by sending up not just one but two satellites into Earth orbit. While the White House worked feverishly to establish NASA as a civilian agency, the military branches were each cooking up their own plans for space. In 1958, the Army forwarded to the White House a plan for a military base on the moon, crewed by soldiers, to both establish a beachhead in space and possibly to site nuclear missiles aimed at the Soviet Union.[1] The Air Force, not to be left out of any lunar bonanza, also put forth a planned military moon base at about the same time,[2] as well as designs for a first-generation spaceplane called Dyna-Soar, similar to the later space shuttle, that would carry a crew, perform spy duties in orbit, and be capable of carrying a variety of nefarious payloads[3]. And there were a number of other schemes smoldering in the military establishments of the US and USSR.

One of the more intriguing, however, had the innocuous name of Project A119, and was called "A Study of Lunar Research Flights."[4] Assembled by the Air Force Special Weapons Center in 1959, this simple unmanned program had one primary objective: to impress the world. While lunar fortresses and bomb-hurling shuttles were intriguing, they would be massive and expensive undertakings, and even in the Shangri-La of Cold War military spending, would turn out to be too rich for the tastes of America's civilian government, which clearly did not see the urgency of militarizing space as quickly as possible. So "A Study of Lunar Research Flights" sought to remedy this via a relatively cheap, easy, and foolproof method of impressing the Communists and their supporters—America would nuke the moon.

The sentiments were stated clearly enough in the abstract:

> Nuclear detonations in the vicinity of the moon are considered in this report along with scientific information which might be obtained from such explosions. The military aspect is aided by investigation of space environment, detection of nuclear device testing, and capability of weapons in space.
>
> A study was conducted of various theories of the moon's structure and origin, and a description of the probable nature of the lunar surface is given. The areas discussed in some detail are optical lunar studies, seismic observations, lunar surface and magnetic fields, plasma and magnetic field effects, and organic matter on the moon.[5]

Easy-peasy—just launch a missile large enough to deliver one of the many nuclear warheads sitting in the US inventory and—*bang*—a bright flash and brand new crater on the moon. Think of all that *science*! Perhaps they would have named the new divot in the lunar surface "Liberty" or "Freedom"—this is not made clear in the study.

What is made clear is an effort to rationalize the project as one with "scientific, military, and political" potential. This mix of objectives is also clearly stated:

> The scientific information which might be obtained from such detonations is one of the major subjects of inquiry of the present work. On the other hand, it is quite clear that certain military objectives would be served since information would be supplied concerning the environment of space, concerning detection of nuclear device testing in space and concerning the capability of nuclear weapons for space warfare.[6]

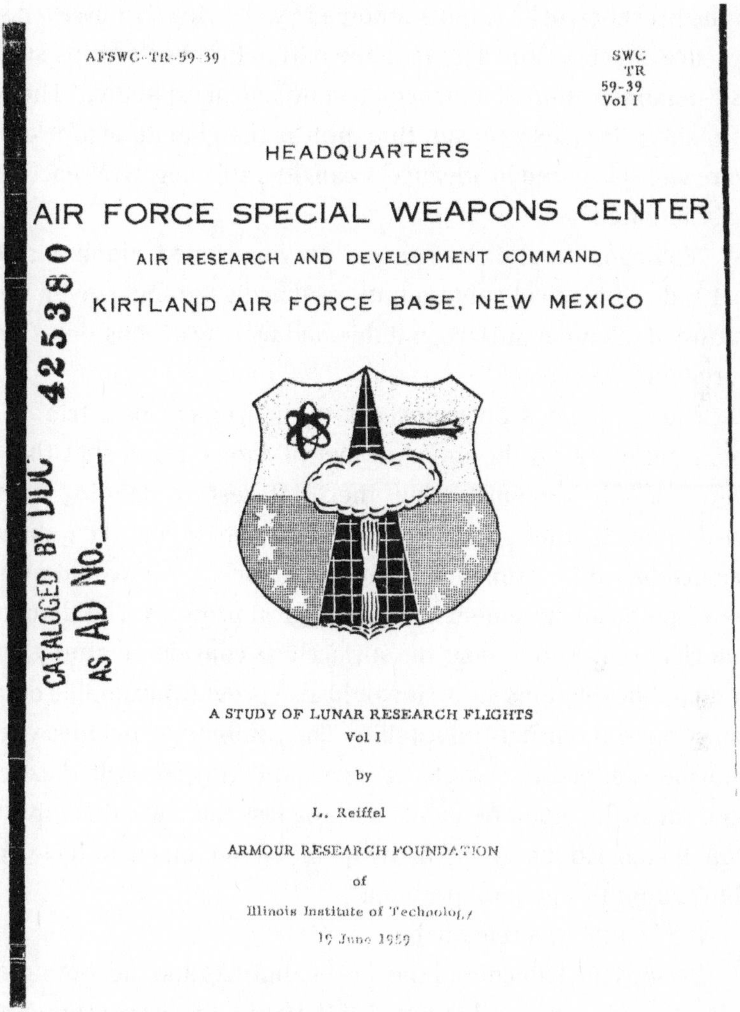

AFSWC-TR-59-39

SWC TR 59-39 Vol I

HEADQUARTERS

AIR FORCE SPECIAL WEAPONS CENTER

AIR RESEARCH AND DEVELOPMENT COMMAND

KIRTLAND AIR FORCE BASE, NEW MEXICO

CATALOGED BY DDC AS AD No. 425380

A STUDY OF LUNAR RESEARCH FLIGHTS

Vol I

by

L. Reiffel

ARMOUR RESEARCH FOUNDATION

of

Illinois Institute of Technology

19 June 1959

Fig. 3.1. An introductory page to the Project A119 study, "A Study of Lunar Research Flights." It was not destined to be flown, but it would have made a briefly thrilling sight on the Earth's natural satellite. Image from DOD.

The mea culpa of the study comes next:

"The political motivations for and against the detonation of a nuclear weapon are equally clear and are, in reality, outside the scope of the present work."[7] This is followed by, "Obviously, however, specific positive effects would accrue to the nation first performing such a feat as a demonstration of advanced technological capability." Then a dose of reality: "It is also certain that, unless the climate of world opinion were well-prepared in advance, a considerable negative reaction could be stimulated."

"Environmental disturbances" are mentioned, along with biological and radiological contamination—hardly damning given the bleak nature of the lunar surface, but deemed worthy of consideration in the burgeoning space age.

The study goes on to discuss the placement of a trio of instrument packages on the moon, a logical choice given that the Soviets were already demonstrating their interest in landing—albeit by hard impact—their own machines on the moon. The delivery of science-based instrumentation to the lunar surface would on its own be a significant statement of technological prowess. The detonation of a nuclear bomb on or near the surface was considered more than just a good public relations stunt; it would also provide invaluable data about the surface of our natural satellite. The summary concludes with, "The enormous effort that would be involved in any controlled experiment on or near the moon demands nothing less than an exhaustive evaluation of suggestions by the many qualified persons who have begun to think about this general problem."[8]

It all sounds so reasonable.

Besides the launches of the Soviet Sputniks and the attempted Luna 1 Russian impactor in January 1959, US intelligence services had also detected radar signals bounced off the moon—to some more fevered minds inside Western intelligence services, clearly a prelude to an invasion of, or at least intervention with, our nearest neighbor in space.[9]

To be fair, the Russians had their own plan—with the equally innocuous name of E-4—to explode a nuclear warhead on the moon and film it to show the world that they had the ability to do so, and, presumably, the ability to land whatever else they wanted to there. This project didn't proceed either; a rare testament to common sense in the early arms race.

Key to the US effort was the notion that the detonation be visible from Earth. This would provide, among the more mundane scientific reasons, a "morale boost" for Americans struggling with the perceived threat of Communism, as well as what was hoped would be a message of intimidation to the Soviet Union. Remember that this was less than a decade after the height of Joseph McCarthy's campaign to stamp out Communism at the most influential levels of the United States via the House Un-American Activities Committee, or HUAC.

Leonard Reiffel, a physicist who later worked on the Apollo program, was the scientific leader of the project, and was joined by a young Carl Sagan, who would later go on to international fame via his involvement with JPL's Viking Mars program. Sagan was also featured in a 1980s television series called *Cosmos*, which would forever enshrine the words "billions and billions" in the global lexicon, referring to the vast numbers of stars and bodies in our galaxy.

Reiffel would later say of the study, "Now it seems ridiculous and unthinkable. . . . But things were remarkably tense back then."[10] He added, "I was told the Air Force was very interested in the possibility of a surprise demonstration explosion, with all its obvious implications for public relations and the Cold War. . . . It was clear the main aim of the proposed detonation was a PR exercise and a show of one-upmanship. The Air Force wanted a mushroom cloud so large it would be visible on Earth. . . . The US was lagging behind in the Space Race."[11]

While weighing the possibilities, the planners concluded that the bomb itself would have to be a small fission weapon (a traditional atomic bomb). A thermonuclear hydrogen bomb was, at that time, too heavy to be carried by any rockets then in the US inventory. But with the deployment of nuclear warhead-carrying missiles due later that year, an atomic detonation on the moon seemed feasible.

The project had its origins in an earlier study produced by the

Armour Research Foundation at the Illinois Institute of Technology, which looked at the effects of nuclear detonations on Earth's environment. In 1958 these studies were expanded to look at similar effects on the moon. Project A119 was further spurred by rumors of the Soviet project, timed to coincide with the anniversary of the famous October Revolution in 1917. An article in the *Pittsburgh Press* said, "The Russians plan to explode a rocket-borne H-bomb on the moon on or about Nov. 7 . . . If that's true—look out! The rocket and its cargo of violence are more likely than not to boomerang"[12]—the implication being that the nuclear maelstrom could likely as not accidentally wipe out a city or region somewhere on Earth instead, a notion that was not lost on the United States, either in regards to the Russian program or their own. "Let us agree," said one scientist, who remained anonymous in the article, "that in view of Sputnik the Russians all of a sudden are all nine-feet-tall. Still, putting Sputnik in orbit was a lot easier from the standpoint of guidance than hitting the moon." Maybe so, but the grapes on the menu of Western scientists sounded a bit sour.

Nevertheless, even famed father of the H-bomb Edward Teller had endorsed such an idea in 1957. The race to incinerate a portion of the moon was on. The bomb of choice was ultimately selected from the US nuclear stockpile—a small 1.7 kiloton device that was about a tenth as powerful as the one used on Hiroshima. This tiny nuke was really little more than a tactical battlefield warhead, but was deemed large enough to send the desired message. To double-down on the likelihood of success, the missile would circle the moon, then detonate near the edge as viewed from Earth, hopefully resulting in not just a flash but a huge cloud of dust that would be illuminated by the sun. One way or another, the explosion, or the results of it, had to be visible from Earth to be an effective deterrent. The dynamics of the cloud of dust and debris are what Sagan had been brought in to study.

A119 was eventually canceled by the Air Force. Reasons cited were public reaction, the risk to the pristine lunar environment, and the possibility of an Earth-endangering misfire.[13]

The moon was safe—for now.

Fig. 3.2. Had the Air Force been successful in launching their lunar nuke, it probably would have ridden aloft on the then just-deploying Atlas missile, seen here in an August 1958 test launch. Image from USAF.

Chapter 4

FRAU IM MOND

The exploration of the cosmos began with a few satellites tossed into Earth orbit. The first, Russia's Sputnik, did little more than orbit the planet and send out a repeated radio beep, but the message was far more profound than the radio signal; to the West, it said, "The USSR is now dominating space . . . the rest of Earth cannot be far behind."

The United States was struggling to put up its own first satellite, called Vanguard. This Navy program, while ultimately successful, failed spectacularly on live TV worldwide in its first attempt, and the United States had to scramble to catch up. That was late 1957, but the notion of exploring the heavens was far older.

In 1929, a new motion picture premiered in Germany before a crowd of almost two thousand people at the UFA-Palast am Zoo cinema in Berlin. The movie was called *Frau im Mond*, or *Woman in the Moon*, and is considered the first serious science fiction film. The plot was pure melodrama, revolving around a mad scientist who envisioned vast reserves of gold on the moon, a greedy American, and a gang of evil henchmen who take over the professor's moon project. An assemblage of these rudely drawn characters is soon launched to the moon aboard a rocket, land on the far side, discover gold, and then struggle to get home. The film was deemed sufficiently realistic that screenings were banned from 1933 until the end of the Second World War by the Nazis, due to the similarities between the rocket technology represented in the movie and the German V-2 rocket program. It could also be seen as presaging the future—if the first rockets in space had some kind of metallic DNA, they would be neither American nor Russian, but rather of presumed Aryan lineage, with bloodlines from Nazi Germany.

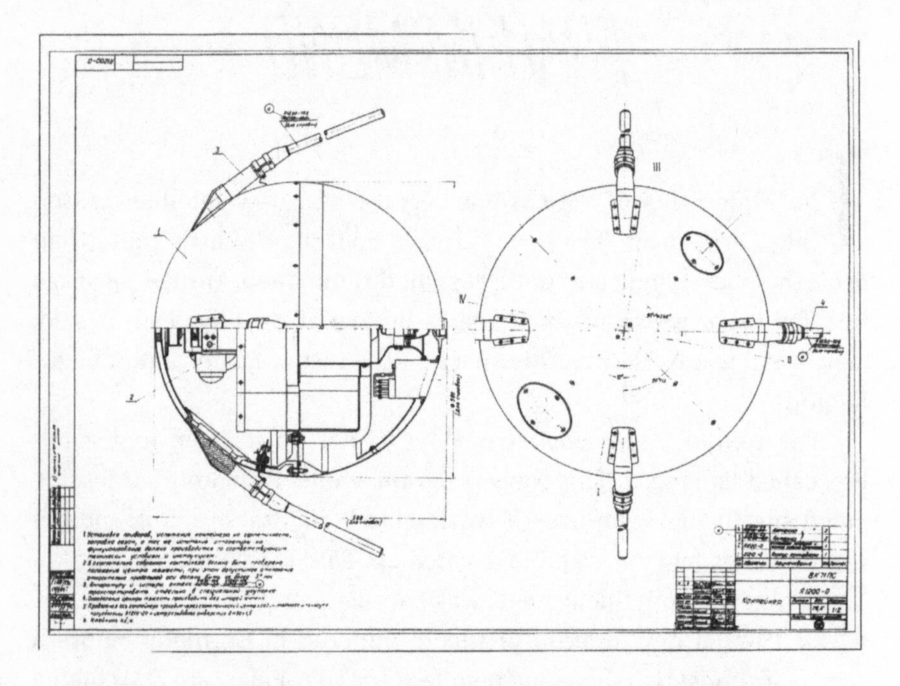

Fig. 4.1. A schematic view of the first Sputnik from 1957. There wasn't much to the metal sphere beyond a battery and a radio transmitter, but it shocked the West when it went by overhead every ninety minutes, beeping merrily away and announcing the accomplishments of the Worker's Paradise of the USSR. Image from NASA.

Then came the Second World War, and with it, the destruction of the German Reich. A nation that might well have attempted the first journeys into space was smashed, leaving a triumphant United States and a shattered Soviet Union to struggle for global influence. Both would build on the engineering accomplishments of the Germans to reach space.

Those accomplishments revolved largely around the V-2 rocket built by German engineer Wernher von Braun and a team of brilliant engineers and technicians he assembled throughout the 1930s. Early on, von Braun's enterprise was purely civilian, but by the mid-1930s had been absorbed by the German military, aimed at producing ballistic missiles that could reach foreign targets robotically. Von Braun's designs proved successful, though incredibly complex and expensive to produce. But the V-2 project was the first foray into successful, mass-produced rocketry, and the results could be seen in countless shattered neighborhoods and bloody shambles throughout the capitals of Europe by 1944. At the end of the war, with the Soviet Red Army advancing from the east and the Allies from the west, von Braun knew he had a decision to make, and concluded that the Americans were the best option to continue his work—and his life. The Russians were not happy with Germany's devastation of their homeland, and treated many of their prisoners accordingly. Von Braun and a couple hundred of his associates quietly surrendered to US forces, and brought much of their hardware, and all of their expertise, to the United States.

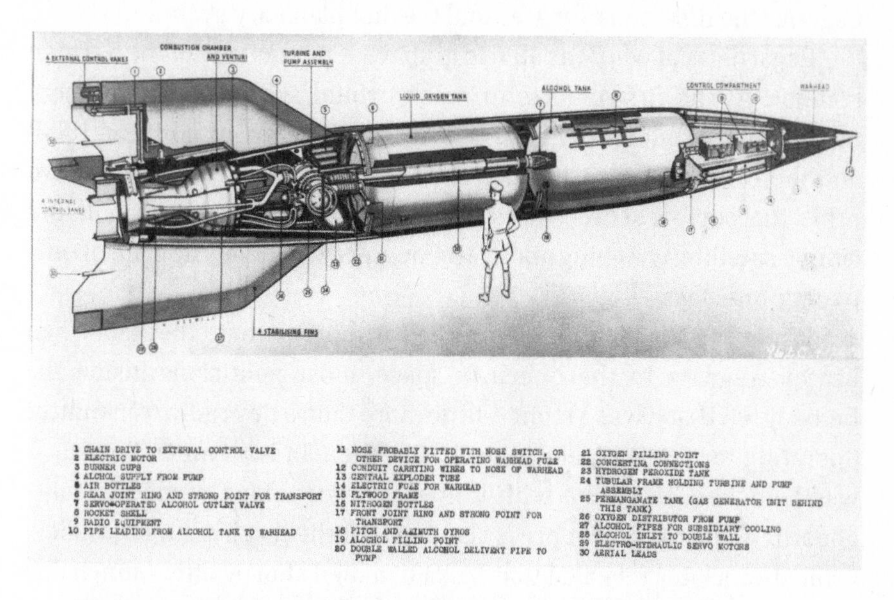

Fig. 4.2. Wernher von Braun's V-2 missile was the father of NASA's earliest rockets, especially the US Army Redstone. Many V-2's came to the US, along with von Braun and a couple hundred of his compatriots, after the surrender of Nazi Germany in the Second World War. Image from US Army.

Along with von Braun and his treasure trove of paperwork, the West grabbed the technology behind the V-2. Despite the fact that the USSR had been part of the Allies during the war, the United States and Great Britain were not eager to share the spoils of German rocketry with Josef Stalin's Russia. The Nazi rocket facilities were quickly scoured by US and British forces, with the US toting away enough hardware to build almost eighty V-2s for experimentation. Every effort was made to spirit away as much of the technology to the West as possible—few trusted the Russians with the Nazi "wonder weapons" of the Second World War.

Nonetheless, by mid-1948, many of the Germans remaining in Soviet-occupied German territory had been relocated to parts of Russia to work in rocket research. While the hardware looked different, there were strong connections between this German expertise and the first generation of large Soviet rockets, including those that launched Sputnik, Yuri Gagarin (the first man in space), and the first planetary probes.[1]

Regardless of who was able to grab what percentage of the German technology, the first objects to reach orbital space were sent there by the Soviet Union. After years of intensive effort in both the USSR and the United States to launch the first artificial satellite into Earth orbit, the Soviets scored in 1957 with Sputnik 1. It was a small and simple satellite, weighing about 184 pounds, with a mission life of only twenty-one days.

While the Western press recoiled in horror that "the Reds" had beaten America to the punch in space, most politicians inside the Beltway felt that it was a stunt—little more than a tiny radio transmitter operating for a few weeks in orbit. But then, in November, Sputnik 2 was launched, this time with a living cargo—a dog named Laika—aboard. While the Soviet press claimed that the dog lived for a week in orbit, it was later revealed that Laika had died shortly after launch due to an overheated capsule.[2] In any event, the message was clear: the Soviets had developed sophisticated launch and control capabilities, and their ability to launch a far heavier spacecraft than the 184-pound Sputnik 1—Sputnik 2 had a mass of over 1,100 pounds—was impressive indeed. This was even more worrisome when one realized that the large Soviet rockets could soon be capable of delivering an atomic warhead across the Atlantic.

Within three months of Sputnik, and after the failure of Vanguard, the United States had followed up with its somewhat more sophisticated Explorer 1 satellite, designed and built at JPL, and flown on a Jupiter-C rocket built by Wernher von Braun and his team in Alabama. The satellite was able to detect cosmic rays, temperature, and micrometeoroid impacts via its compact instrument package—a much higher science return than the first Sputnik.

But the launching of large and heavy spacecraft, hurled into orbit by big rockets, continued to be a Soviet game. The rapid-fire accomplishments of their first satellites were followed by manned flights just a few years later—Yuri Gagarin's single orbit of Earth in 1961, and the first spacewalk by Alexey Leonov in 1965, which demonstrated a capability that continued to exceed that of the United States. America's Mercury flights followed close on the heels of their Soviet counterparts, but until the flights of the Gemini program began rendezvous and docking tests in the mid-1960s, human spaceflight was dominated by the Russians.

And so it was with robotic planetary exploration in those early years. While the two superpowers were very publicly competing for human spaceflight firsts, a less visible, but still hotly contested, competition to explore the planets was underway. Long before US president John Kennedy issued the lunar landing challenge to the Soviet Union in 1961, space scientists in both countries had been longing to send robotic probes to the moon and beyond. Satellites in orbit were a major accomplishment—and necessary first step—but for the scientific academies and space agencies of the US and USSR, reaching and exploring other planetary bodies was the primary goal for unmanned spacecraft.

And here, too, the Russians proved superior—for a time.

In rapid-fire order after the first Sputniks orbited the planet, a series of Soviet probes called Luna were sent toward the moon. Luna 1 successfully broke free of Earth orbit and headed toward lunar orbit in January 1959, but did not achieve its intended goal of impacting on the lunar surface due to a rocket malfunction shortly after launch. The probe missed the moon by 3,660 miles, eventually entering an unintended orbit around the sun. But it came close to success.

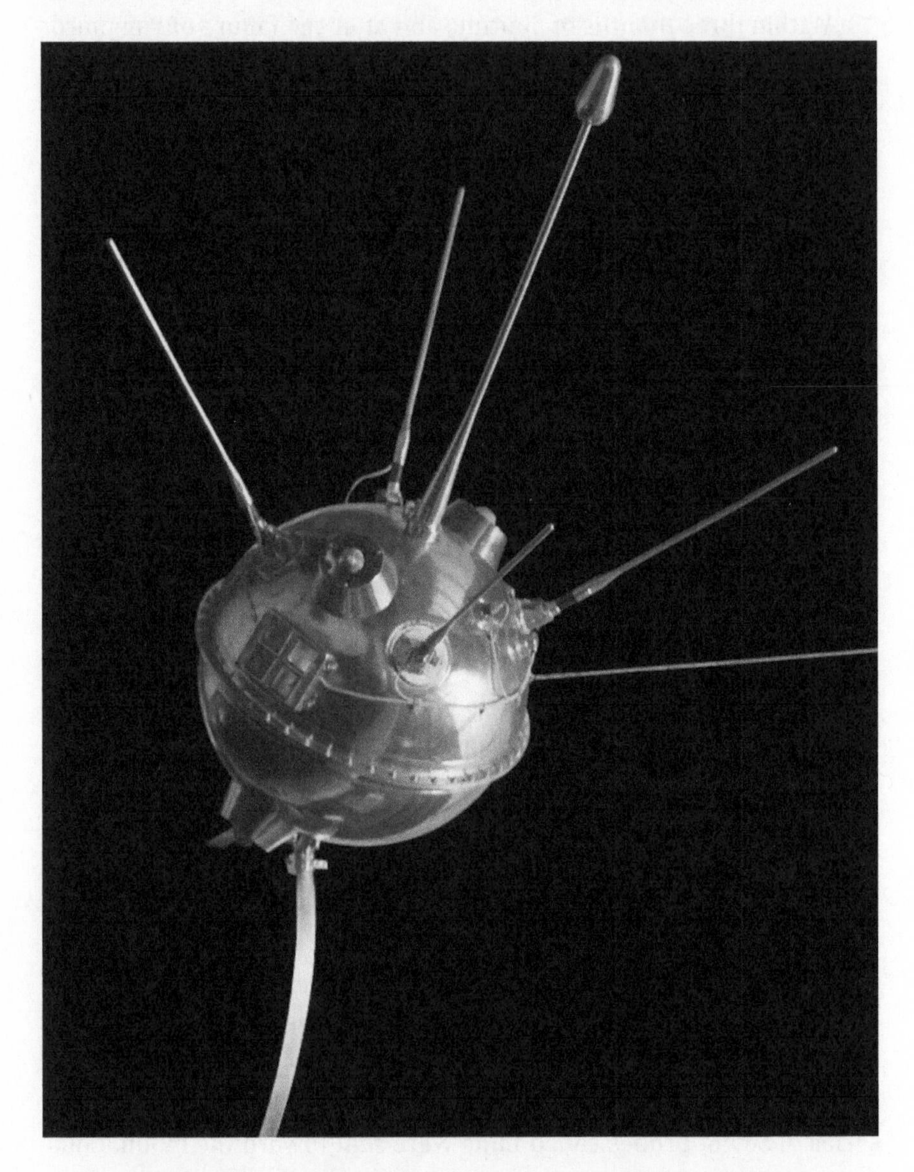

Fig. 4.3. Luna 1 prior to launch—it ended up orbiting the sun instead of impacting the moon, but held the distinction of being the first attempted deep-space probe. Image from NASA.

In September of the same year, Luna 2 succeeded in its suicidal goal of impacting the lunar surface just off the visual center of the moon as seen from Earth—the first spacecraft to reach the surface of that world. The probe carried a basic suite of scientific instruments that were similar to the Earth-orbiting satellites of the same era: radiation detection equipment, micrometeoroid impact detectors, and a magnetometer for detecting magnetic fields. Perhaps more patriotically, Luna 2 also carried small liquid-filled metal spheres, looking like metallic soccer balls, with the emblem of the Soviet Union repeated across their surface. These were designed to shatter upon impact, scattering the CCCP-engraved shards across the lunar surface and permanently marking the site of the first man-made item to arrive on the moon. It has been suggested that, despite the political value of "claiming" the lunar surface, the Soviet souvenirs probably vaporized upon impact.[3] A replica of one of these spheres was later presented to US president Dwight Eisenhower by the Soviet leader Nikita Khrushchev as a gesture of goodwill . . . and a sign of continuing Soviet superiority.

It is worth noting that this first success followed four failed attempts. In a trend that would continue for decades, refined development was traded for quantity. The Soviets hurled probes at the moon, and later at planets, with astonishing frequency, achieving a small number of successes as they proceeded. But succeed they did, and this space-age strategy takes nothing away from their achievements in planetary exploration.

Just a month later, in October, Luna 3 was launched. Its mission was decidedly different: this was a flyby probe, designed to photograph the lunar surface from a distance of about 3,800 miles. On October 7, the simple onboard electronic photocell detected sufficient sunlight bouncing off the brightly lit lunar surface and Luna 3's cameras were activated. This in itself was impressive—the Lunas were the first spacecraft to carry cameras beyond Earth orbit, and to take photos by remote control and transmit them back to Earth. And, in a major propaganda coup, Luna 3 did accomplish one more amazing feat—

the images it sent back to Moscow were of the *far* side of the moon, never before seen from Earth. The moon is what is known as a "tidally locked" body, rotating at the same rate at which it orbits the Earth. As a result, we always see the same side. Besides reaching the moon with a robotic machine, the Russians had managed to send back seventeen images from the unseen backside of the moon, covering roughly 70 percent of that hemisphere.

The images themselves were taken by a traditional film camera that operated automatically once the bright lunar surface was within range. A strip of 35 mm photographic film was held on spools, and slowly inched its way past the two camera lenses—both telephotos—as the shutter snapped. For the next forty minutes, the far side of the moon was photographed. The film then proceeded into chemical baths to process it.

The film was dried onboard, then sent to an electronic scanner. This used a bright spot created by a cathode-ray tube—not unlike the earliest TV sets—which shone through the developed film. The resulting image was received by a photosensitive electronic device called a photomultiplier tube. This amplified signal was then transmitted back to Earth via radio, using frequency modulation—essentially FM radio—to indicate varieties of brightness (a technology similar to that used by fax machines later in the twentieth century). The spacecraft finally transmitted these images after it looped the moon on its single pass and was headed back toward the Earth. The film itself, which was designed to be insensitive to the radiation in space that might have otherwise caused it to be prematurely exposed, is said to have been recovered from US spy balloons flown over the Soviet Union a few years before. A few of these were shot down by Soviet fighter aircraft and were quickly disassembled and salvaged once found on the ground.[4] So, in the final analysis, the United States was upstaged by images taken on its own film.

The low-resolution images of the lunar surface were surprising to the scientists studying them. While the nearside of the moon, that visible from Earth, is mottled with both light and dark areas, the backside was not. The large swaths of dark material on the nearside are called *mare* (Latin for "seas"), and are the result of ancient lava flows. The backside appeared to contain little of this material and was far more severely cratered—the huge number of impact scars had never been covered by subsequent lava flows. It looked like an adolescent with severely troubled skin. The fuzzy images also hinted at a vast impact crater on the lunar south pole, later named the South Pole-Aitkin Basin, one of the largest impact craters in the solar system at 1,616 miles across.

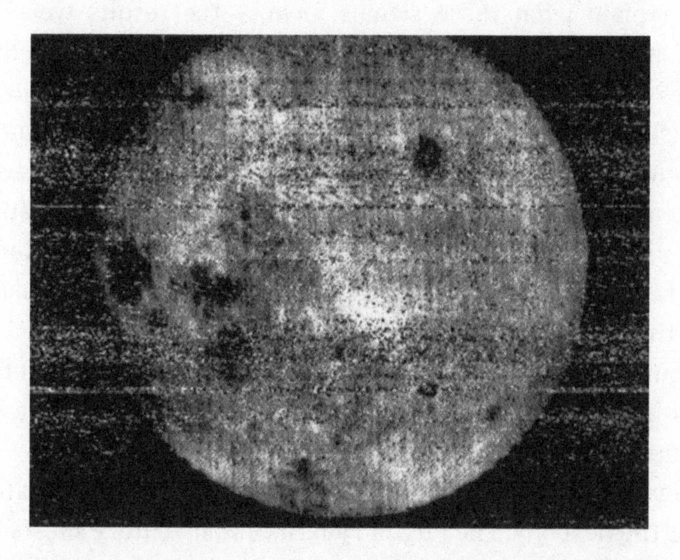

Fig. 4.4. The first image ever returned of the far side of the moon, which can only be seen via spacecraft. This shot from the USSR's Luna 3 was a sensation in 1959. Image from NASA.

A year later the Soviets published an atlas of the lunar farside. These images would not be updated until the manned Apollo 8 spacecraft orbited the moon in 1968, and the Russian book was an instant collectable. Score one more victory for the USSR.

As a side note, the United States was sufficiently alarmed by the early Soviet prowess with its lunar program to attempt a bit of intelligence-inspired misbehavior in 1959 or 1960—the exact date remains unclear. By 1959, Russian lunar probes had enjoyed some success—while the first one had missed the moon, Luna 2 had been deliberately crashed there and Luna 3 had returned those irritating (to the US) images from the lunar farside. America's space program looked like it was standing still. In a brash adventure, a CIA team of four planned to heist a Soviet moon probe for examination—the US wanted whatever information it could get from the Russian program.[5]

After months of intercepting telemetry from Soviet launches and the lunar probes—without the benefit of any Russian manuals that would explain what these signals meant—the results were unsatisfying, so this more direct action was approved. US intelligence agencies knew that the Soviets were planning a multi-country tour showcasing their technological might, and they were fairly certain that an unlaunched lunar probe would be among the bounty on display.

One night, sometime in 1959 or 1960 (the exact dates are still classified), US agents snuck unobserved into the exhibition hall after it had closed for the day. An examination of the spacecraft on display confirmed that this was not a model, but the real thing—a complete, functional Luna probe, sitting alongside a Sputnik. Bingo—it was time to plan for a closer examination. The agents crept back out of the exhibition, emptyhanded but mulling plans for a major heist.

After extensive preparation, the agents waited for the exhibit to move to the next city. The Russian guards had inventory sheets for the crated exhibits, but they were not in the habit of examining each one as it was loaded onto and off of the trains and trucks that transported them. As reported, CIA agents tailed the truck onto which the Luna exhibit had been loaded, stopped it somewhere along a lonely patch of highway, took the local driver to a hotel for the night (he was clearly not an ardent Communist), covered the truck with a tarp that disguised its appearance, and drove it and the precious cargo to a nearby salvage yard recruited for the purpose.[6]

After waiting for the better part of an hour to make sure their subterfuge had not been detected, the agents set to carefully removing the lid of the crate, making sure not to leave any markings on the wood that might signal their activities upon later inspection. Others prepared cameras to record the dissection. Like a scene from a Tom Clancy novel, they descended from above the crate via rope ladders and began to disassemble the Luna probe while it was still sitting inside—nobody knew for certain that there was not some kind of telltale device affixed to the interior, or even possibly a booby trap.

Upon disassembly, the agents noted that, while the small rocket engine had been removed from the probe, the fuel tanks and associated plumbing were still intact, giving them direct evidence of its capacity. On the other end of the Luna, the stealthy project became more complicated—the machinery was affixed with a plastic seal embossed with a Soviet emblem, right where they would need to disassemble it. Undeterred, they proceeded, and the pieces of the now-destroyed seal were sent out for other CIA operatives to reproduce a convincing replica that would replace it when their examination was complete. These folks knew how to multitask.

After working all night and carefully documenting each component that they could access, the agents reassembled the spacecraft as dawn approached. The last component to be replaced was the replica plastic seal created by their co-conspirators. The truck's original driver was brought back by 5:00 a.m., and delivered the cargo to the railyard before the Soviet guards showed up at 7:00. Off went the Luna to its next public showing, with nobody any the wiser.

Knowing the mass and configuration of the Luna probe allowed intelligence analysts to reverse engineer subsequent launches (of which, as we have seen, there were many), revealing among other things the true lifting capacity of the Soviet boosters. And so it was that, early in the space age, a discreet peek inside the Russian lunar program gave a leg up to the West.[7]

Despite the success of this intelligence gathering, the United States was taking a somewhat different approach toward similar goals. The first American probes intended to travel beyond Earth orbit were called Pioneer, and the earliest of these were intended to be lunar orbiters with a mission profile similar to the orbiting Luna probes.

The Pioneers also suffered from initial failures, due mostly to rocket malfunctions and navigation equipment issues; Pioneer 4 did succeed in flying past the moon in 1959, but its scientific instruments failed to activate. One additional Pioneer, intended to fly past Venus, was delayed at launch and, having missed its launch window, was rerouted to explore interplanetary space inside Earth orbit. While the other Pioneer missions struggled to reach the moon, Pioneer 5 succeeded with its repurposed mission—the reconnaissance of interplanetary space—in 1960. But the moon remained elusive.

Another American program, called Mariner, began to reconnoiter the solar system, with Mariner 2 successfully flying past Venus in 1962—a first—and Mariner 4 dashing past Mars in 1965.

But soon, American spacecraft would reach the moon, leaving its own kinetic trademark by crashing into that body.

CHAPTER 5

SUICIDE ON THE MOON

The United States' efforts to reach the moon with a functioning spacecraft were frustrated until 1964, a full five years after the Soviet triumph of Luna 2. The US Ranger program started development at JPL in 1959, shortly after the first Pioneer launches. The Ranger program would proceed in phases, each with specific goals. The first two Rangers flew in 1961 to test the spacecraft in Earth orbit, but both failed to achieve their objectives. This was more a case of rocket failure than of the Rangers themselves, though, due to a separate malfunction, Ranger 1 did try to deploy its solar panels before launch and while still inside the protective launch fairing. But the real culprit was the upper stage of the Atlas rocket, called the Agena—in both of these first Ranger launches, this second stage failed to operate properly (Agena failures would continue to frustrate NASA during the Gemini program years later), and placed the Rangers into unstable, low orbits. The probes simply did not have time to perform their appointed tasks.

With the first orbital flights written off, NASA went on to directly assault the moon. The Soviets had proved to be capable of reaching our natural satellite, and by 1961 they had orbited a man around the Earth. NASA's string of disappointments was growing, and a success in robotic spaceflight was needed as badly as one in manned spaceflight.

There was another factor in this thinking. In May of 1961, President John F. Kennedy announced his decision to a stunned Congress that he was putting America on a course to "landing a man on the moon and returning him safely to the Earth."[1] This proclamation was cemented into the public mind by another speech—this time to the public—at Rice University in the summer of 1962.[2] But before humans could be sent there, the machines would have to go first, to reconnoiter the surface of that world and give us some inkling of what we were up against. The lunar robotic program switched into a higher gear.

The next batch of Rangers were designed to fly to the moon and transmit television images as they got closer and closer, right up to the point of impact. They were, in effect, little kamikaze probes, intended to crash. Odd as this may seem today, at the time it was deemed the most reliable way to get a series of ever-closer looks at the lunar surface during the final ten minutes of the doomed spacecraft's life. By continuing to transmit until the final moment, surface features as small as a dinner plate could, in theory, be imaged by the TV cameras.

The Rangers were larger than the Pioneers, and almost ten times as heavy. Rangers 3, 4, and 5, the so called "Block II" versions, were more ambitious versions of their predecessors, and were actually two spacecraft. The main structure consisted of a five-foot octagonal base, with a cone-shaped tower rising about eight feet above it, which housed the TV camera, along with instruments to measure the dynamics of the environment between the Earth and the moon. Two solar panels swung out from the sides of the base to provide power. But there was more to this second-generation Ranger. At the top of the instrument tower rode a two-foot-wide ball with a lunar seismometer inside, surrounded by a cushioning jacket of water. All this was covered with a thick layer of balsa wood.

This wooden sphere was intended to protect its payload for a "semi-soft" landing—aka a controlled crash—into the lunar surface. Wood and water—two of the five ancient Chinese elements—were the best protection for a hard landing that could be devised at the time. It seemed like a good idea . . . but everything else would have to work properly for it to be tested in action. Ironically—for this package failed every time it was tried—the balsa wood sphere was painted like a circus ball, the kind seals used to balance on their noses. It was, inadvertently, an appropriate touch.

Fig. 5.1. Ranger 2 with its innovative circus-ball seismic probe mounted at the top. It never worked. Image from NASA.

Ranger 3 was launched in January 1962 but due to a series of rocket malfunctions missed the moon by 22,000 miles. This was six times further from the moon than Luna 1's passage two years before, and yet another embarrassment. In April, Ranger 4 followed, but a failure of the primitive onboard computing system caused it to crash on the lunar farside, out of radio contact. It was, however, the first US spacecraft to reach the surface of another celestial body, albeit destructively.

In October, Ranger 5 headed moonward, but missed the moon by 450 miles. The trend was unsettling, but at least NASA was getting closer.

At this point, a frustrated NASA dictated a redesign of the Ranger spacecraft, removing the "hard lander" wood-enclosed lunar seismometer. The third iteration of Ranger would be designed to simply crash into the damned moon—if it could find its way there.

The improved Ranger had six TV cameras—two with wide-angle lenses, and four more with telescopic lenses. It was loaded with instrumentation to assure success. As Edgar Cortright, then the deputy director for Space Science and Applications at NASA headquarters, put it, "It really was tense, because we were investigated on every one of those failures, and as it got worse and worse, we'd spend weeks in front of Congressional committees, explaining it, and I had a lot of that. . . . I had to do all of it."[3]

So they changed not just the spacecraft's design, but how they went about building it as well. "In the beginning, you see, we didn't want to contaminate the moon, so everything had to be sterilized," Cortright said. "JPL tried to do it by heat sterilization, and all the wiring ended up like spaghetti, so that failed."[4] It was a burned, melted mess. "We gave up on that and said, 'They really don't have to be sterilized. They just have to be clean.' So we cleaned it with a gas [and] . . . liquids that were disinfectants, germicide-type things, and tried to keep it as clean as possible." But there was one more failure in store, and that would be Ranger 6. It had gotten so bad at that point, Cortright recalled, that at "one time the program was called 'shoot and hope.' But finally, 7, 8, and 9 worked."

Ranger 6 reached the moon, but experienced a failure of its TV system and no images were received. But this was an improvement over missing the moon entirely, and with Ranger 7 NASA hit pay dirt. Launched in July 1964, the spacecraft successfully transmitted 4,300 images during the last seventeen minutes of its short life. The impact site was in the general vicinity of the 1969 Apollo 12 landing site.

The increasingly closer images taken in those last few moments revealed that the large craters evident from Soviet probes and from telescopic images taken from Earth simply scaled to smaller and smaller ones as the probe got closer to the moon . . . there were craters within craters within craters. It was a rougher surface than most had suspected, and this was valuable data for those planning the Apollo lunar landings. There was more to come.

In February 1965, Ranger 8 swept across the lunar nearside, crashing near the eventual Apollo 11 landing site in the Sea of Tranquility. The 7,000 images returned showed the same general surface as had been seen with Ranger 7—a devastated, densely cratered swath.

A month later, Ranger 9, the last flight of the project, sent back 5,800 images of another part of the lunar surface, this time lit at a more extreme angle, with the sun lower in the lunar sky. This provided much more relief in the surface texture, and the rough terrain was easier to discern—and far scarier looking.

Fig. 5.2. The final redesign for Ranger resulted in a string of successful flights, providing NASA with its first extreme close-ups of the lunar surface they intended to land humans on in a few short years. Image from NASA.

The Ranger program, after a troubled beginning, had finally scored big. As William Pickering, JPL's director at the time, later said, "After the Ranger [program] was so successful and was such a good public success—because we were able to do live television of it flying into the moon—after that, we didn't really have much trouble with Congress."[5] JPL had redeemed itself.

But NASA was not yet done with the pre-Apollo unmanned reconnaissance of the moon. From 1966 through 1967, it sent another series of spacecraft, called the Lunar Orbiters, to photograph the lunar surface from stable orbits. The Lunar Orbiters mapped most of the moon's surface with a maximum image resolution of about three feet.[6] The images were not as detailed as the final ones from the Rangers, but they provided a good overall map of lunar terrain on both the nearside *and* the farside, which can only be seen by spacecraft.

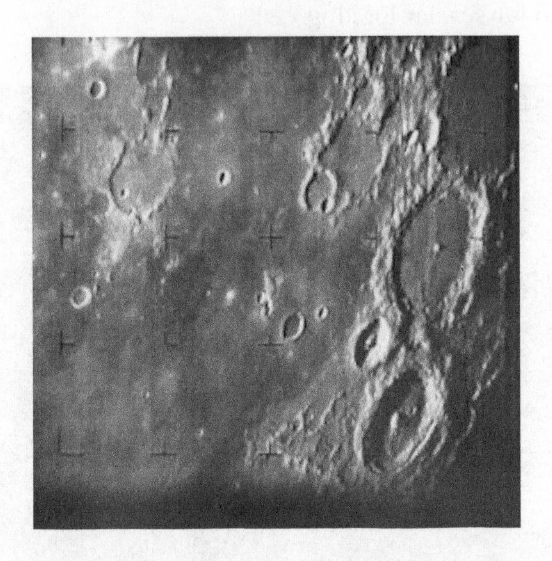

Fig. 5.3. An image from Ranger 7, taken seventeen minutes before it impacted into Oceanus Procellarum, which would be the landing site of Apollo 12 a few years later. Image from NASA.

Lunar Orbiters 1, 2, and 3 focused on planned Apollo landing sites, providing useful imaging of each. In this way, decisions could be made about the general roughness of the terrain and promising geological features—craters punched through otherwise smooth, basaltic terrain caused by lava flows could reveal samples of lunar bedrock that could be collected, for example. Similarly, mountains and rilles would allow astronauts to collect samples differentiated from the lunar plains. The Lunar Orbiters also provided the imagery that would be used to assemble the maps that early Apollo missions would use for navigating to a landing.

With the equatorial regions well charted, the remaining two Lunar

Orbiters, 4 and 5, were sent into higher-inclination orbits that allowed them to pass over the moon's poles, and by the end of the program nearly 99 percent of the surface had been imaged. It was a staggering achievement after the struggles of the Ranger program, and it showed how quickly NASA and JPL were advancing once the space race was in full tilt.

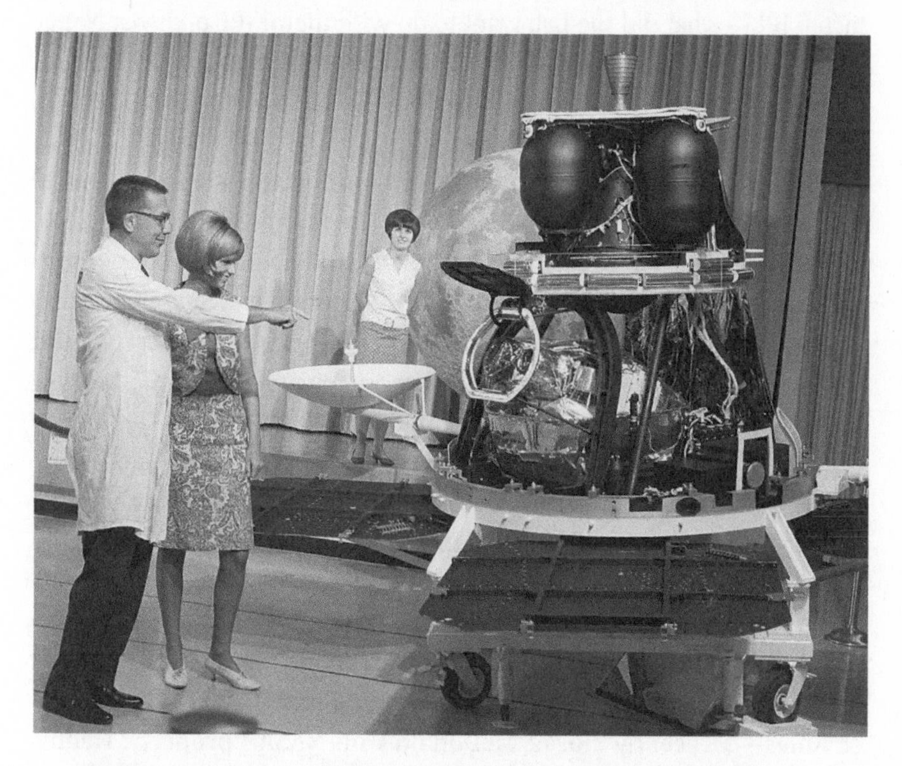

Fig. 5.4. In a rather posed-looking photo, a NASA technician explains Lunar Orbiter 1 to a painfully hip listener at a NASA press event. Lab coats always lent authority to scientific outreach events . . . at least in the 1960s. Image from NASA.

In an interesting side note, the Lunar Orbiter image data lives on today. When the program was flying, the images the probes returned were stored on then state-of-the-art two-inch magnetic tape. Once the maps had been created and the Apollo landings made, these tapes were sent, along with much of the data from other lunar probes and the Apollo missions, to the National Archives in Maryland for storage. For two decades, the 1,500 tape reels sat in obscurity, unwanted and unneeded. Then, in 1986, someone at the archives decided it was time to clean house—NASA had no further use for this material, and conventional practice would have been to send them to a shredder. But before proceeding they wisely contacted JPL—what did the Lab want to do with them? JPL archivist Nancy Evans felt that they might still be valuable, even if simply as a historical reference. "Do not destroy those tapes," she said when asked, later adding, "I could not morally get rid of this stuff."[7]

The tapes sat at JPL for a few years, until Evans scrounged some money to find tape machines that would run them—all but impossible to locate after two decades. These were Ampex-manufactured two-inch quadruplex videotape players, long since discontinued. Spare parts and documentation came from caches of government surplus materials. But the specialized hardware required to resolve images from the tapes—called demodulators—were long gone, and there was no money to create new ones. The tapes found a new home in Evans's garage.

In the mid-1990s, a researcher at NASA headquarters took up the trail again and started asking around to see if he could find some help to work with the treasure trove of lunar images. It took the better part of a decade, but by 2007 the tapes had been moved to surplus space at NASA's Ames Research Center in Sunnyvale, California. They sat for a time in storage, and were then moved to newly available space at Ames—a recently closed McDonald's on NASA's property. Dennis Wingo and Keith Cowing, both active in private space efforts and space media, raised funds to set up a restoration and conversion facility beside the unused deep fryers and cash registers that remained in the building. They took to calling it "McMoon's."[8]

A corps of university students and some retired Ampex employees were recruited to the effort, and the results were quick to arrive. After months of some trial and error, due to the fact that the included documentation from the fast-moving Lunar Orbiter program was incomplete, stunning black-and-white images began to appear at McMoon's. Whereas the original images from the 1960s had been fuzzy and streaked with interference—the shots released to the public had been taken from video monitors on 35 mm film—the newly restored digitally enhanced imagery was astounding. The visual range was about four times the originals, and the resolution doubled. After countless hours and hundreds of thousands of dollars, the team had something to be truly proud of—not simply a piece of history, but a priceless visual record of the moon as it appeared in the mid-1960s. Lunar scientists were quick to compare the images to new ones taken by twenty-first-century lunar orbiting spacecraft, enhancing their understanding of decades of changes on the lunar surface, small though they might be.

But back to the 1960s: Humans had not yet landed on the moon, and there was one more task for the robotic explorers before they could do so—sending robots to make sure the surface was safe.

Chapter 6

FLASH FORWARD: THE LUNAR FLASHLIGHT

Despite the success of the Apollo missions with their exploration of the lunar surface, there is still much to be learned about the moon. And with a new emphasis on the moon by the Trump administration, making the moon NASA's prime target for human exploration over the next decade, it is imperative to learn more about what resources might be found there. Planetary scientists and mission planners have talked about harvesting resources from the moon for decades—a process called In-Situ Resource Utilization, or ISRU. Analysis of the Apollo moon rocks showed that there are metals in them—aluminum, titanium, and more—as well as a surprising amount of water. More recently, hints of larger deposits of pure water ice have been spotted at the lunar poles. The poles are particularly promising because craters there can be in permanent darkness, and water deposited from comets and asteroids (both bountiful sources of water ice from the early solar system) can remain in an eternally frozen state. And where there is water, simple processes can create rocket fuel (hydrogen and oxygen, from splitting the water), as well as breathable oxygen and, of course, drinkable water. NASA's got a plan for that.

Lunar Flashlight is a long-planned, low-cost lunar orbiter that would launch on an early flight of NASA's new Space Launch System (SLS) as an "auxiliary payload," in essence hitchhiking a ride with the larger primary payload atop the giant new rocket.[1] The probe uses a standardized satellite design called a 6U CubeSat, and is about as big as an oversized Amazon Echo speaker, but square on the sides. Tiny and light, the probe will use a very small rocket motor to go out to lunar orbit, and once there it can be maneuvered independently. Lunar Flashlight has been developed by JPL, NASA's Marshall Spaceflight Center, and the University of California at Los Angeles (UCLA).

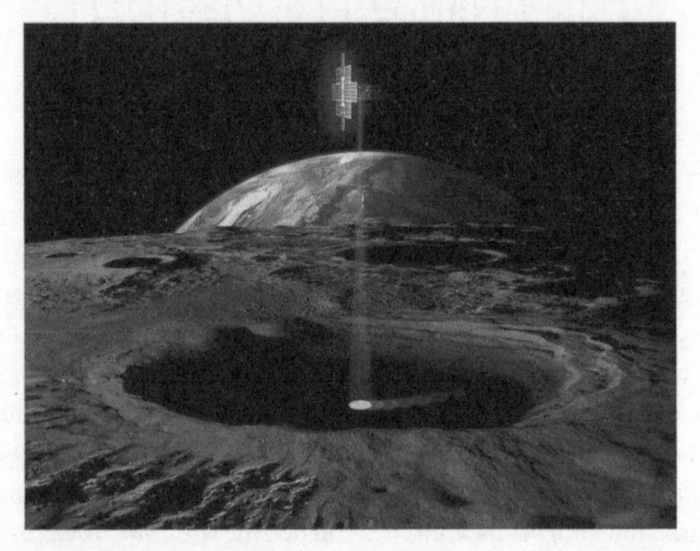

Fig. 6.1. The Lunar Flashlight is a planned mission that will utilize the affordable CubeSat form factor to send a small satellite to the moon, where it would identify water deposits frozen in permanently dark cold traps. The mission has been delayed many times, but it may fly on SLS once it's operational. Image from NASA.

The spacecraft's primary purpose is to detect water ice deposits at the lunar poles. Previous missions, such as NASA's Lunar Crater Observation and Sensing Satellite (LCROSS), their Lunar Reconnaissance Orbiter (LRO), and India's Chandrayaan-1 orbiter, all detected water on the moon in trace amounts, adding to the conviction first formed with the analysis of the Apollo lunar samples that there might be more water there than we had thought. But this data has primarily been gleaned from thermal mapping, and the evidence is incomplete. That's where the Lunar Flashlight mission comes in.

Lunar Flashlight, orbiting high above the poles, will use an infrared laser to shine light into the darkest polar regions, and the resulting illumination of the surface will be analyzed by an onboard spectrometer. That analysis will tell researchers what lies in the permanently shadowed craters, and any water ice there should be immediately apparent.

Because the technology involved has already been flight-proven, it's an extremely cost-effective mission to fly. The recent launch of the Mars InSight lander includes two CubeSats to set up an experimental orbital data relay—the first CubeSats flown to another planet—and CubeSats have been used in Earth orbit for decades. Similar spectrometers have also flown on a bevy of missions, including NASA's Moon Mineralogy Mapper, and are proven technology.

If Lunar Flashlight is successful in its mission, the implications could be vast. Despite a general downward trajectory in the cost of launching supplies into space, it still costs many thousands of dollars to send even a single gallon of water into orbit—even at SpaceX's current bargain launch costs, about $10,300[2]—and the same holds true for rocket fuel for use in space, as well as compressed oxygen for breathing and other uses. The ability to harvest these resources from raw materials found on the moon would be a game changer for the future of human exploration beyond Earth orbit.

Chapter 7

MARS IN THE CROSSHAIRS

Mars died in a single day.

To be more specific, the classical vision of Mars as a place colder than, yet not unlike Earth, vanished suddenly. Prior to July 15, 1965, little was known of the red planet. NASA headquarters had been almost single-mindedly focused on the manned spaceflight effort to land a man on the moon, and JPL had enjoyed little success with their moon-bound robots until 1964. One repurposed Ranger, renamed Mariner 2, had succeeded in flying past Venus in 1962. But the Lab was also quite interested in Mars, and a newly designed Mariner spacecraft was planned for a 1965 launch toward that planet—but, given JPL's track record at the time, would it work?

Thankfully, this would be just a flyby mission, so the engineering around the complexities of trying to orbit or land on the planet would not be necessary for years in the future. Far too little was known about Mars to make responsible decisions for landing there. Nobody had a true measurement of Mars's atmosphere, which would be needed to facilitate a landing. Estimates from the 1950s were that it was about one-tenth that of Earth's, and, as it turned out, these assessments were off by another factor of ten. But at the time nobody knew that. Mars was a mystery, a wavering red orb in the eyepiece of a telescope.

There was also much disagreement about what might lie on the surface of Mars, beyond red sand and rocks. Since the advent of large telescopes, astronomers had assiduously charted the Martian surface with uncertain results, since at its closest the planet was still well over

30 million miles from Earth-bound observatories. While the Victorian-era notion of artificially created canals had been long dismissed by most astronomers, strange "waves of darkening" were seen seasonally across the planet, and seemed to occur in repeating regions. Many astronomers thought them to be sandstorms, but some maintained that these might be vast, seasonal growths of plants such as lichens, which could survive in the carbon dioxide–rich atmosphere of the planet. Mars was still so uncertain a place that in the early 1960s JPL was still using telescopic charts from the early twentieth century to plan its assault on Mars with Mariners 3 and 4. There had been no real attempt at updating these maps because the quality of telescopic observations had been fixed since at least 1948, when the last large telescope, Mount Palomar's 200-inch instrument, had become operational.

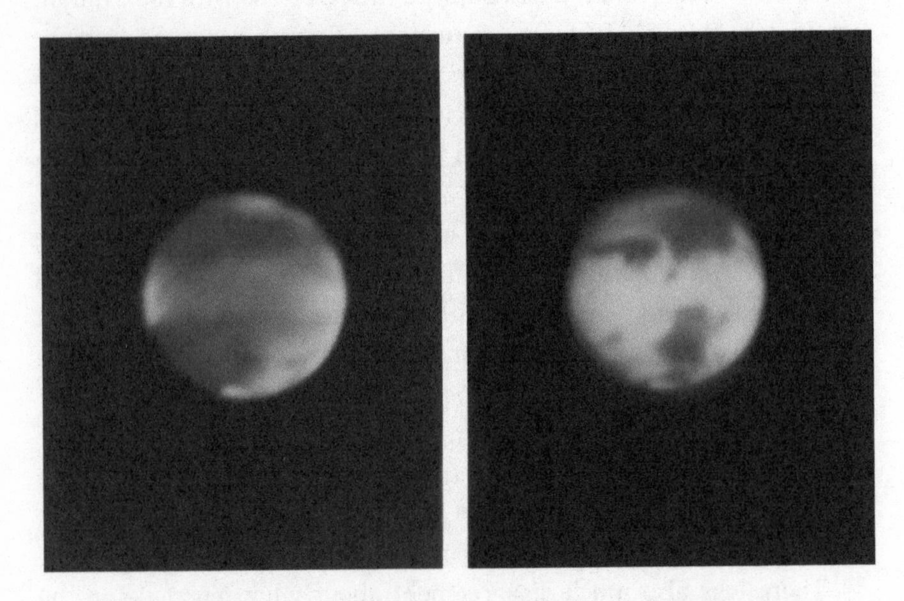

Fig. 7.1. Both hemispheres of Mars as seen through Mount Palomar's 200-inch telescope. This was about as good as it got until Mariner 4 flew past the planet in 1965. It's easy to understand how the imagination of people—and even scientists (yes, they are people too)—backfilled the lack of detail with their imaginations. Image from NASA.

There were even a few romantics that still clung to Victorian notions that there might be canals on Mars, possibly natural channels, maybe even constructed by intelligent beings. These lines were charted across the Martian surface by astronomers such as Percival Lowell at the turn of the twentieth century, and Lowell's interpretation was that lines must have been created by intelligent beings. But different astronomers saw different things, and it has recently been suggested that Lowell and others were actually mapping reflections of the capillaries in their eager eyeballs upon the eyepiece.[1]

More realistic minds saw the planet as a chilly landscape of red plains, possibly dotted by verdant forests of Martian trees or moss. But on a July day in 1965, when Mariner 4 was about to fly past and then behind the red planet, anything was still possible. The minds in JPL's mission control were concentrated on just one thing: get back the photos and the scientific data. We would know the truth, one way or another, at last.

But first, some delightful history: Mars has a colorful past. During ancient times, Mars was not a *place*, but rather an idea. The ancients saw the planet as a god, usually one with unfortunate associations. To the Greeks, the blood-red star was Ares, the god of mayhem; a lusty, violent frat boy who flirted with other men's wives and tipped over the smoldering barbecue. By Roman times, the god was called Mars, and while his habits were similar to that of the Greek deity, he was seen in more positive terms—these were Romans, after all. Violence was *cool*. The ancient Egyptians and Chinese saw Mars in broadly similar terms, often associated with fire, blood, and war.

Then, when telescopes emerged as investigative tools, Mars came to be known as a place. It was another world, as famously outlined by H. G. Wells in his seminal science fiction novel *The War of the Worlds*. Thankfully, Wells's vision of plundering, murderous Martian overlords was not to be. But many others, noted astronomers among them, still thought it possible that the planet might harbor intelligent life.

Fig. 7.2. The Greeks called Mars Ares, and saw him as a troubled youth—violent and warlike, but not very smart, and prone to getting himself in trouble, much like your wayward son-in-law. Image from Wikipedia/Jastrow.

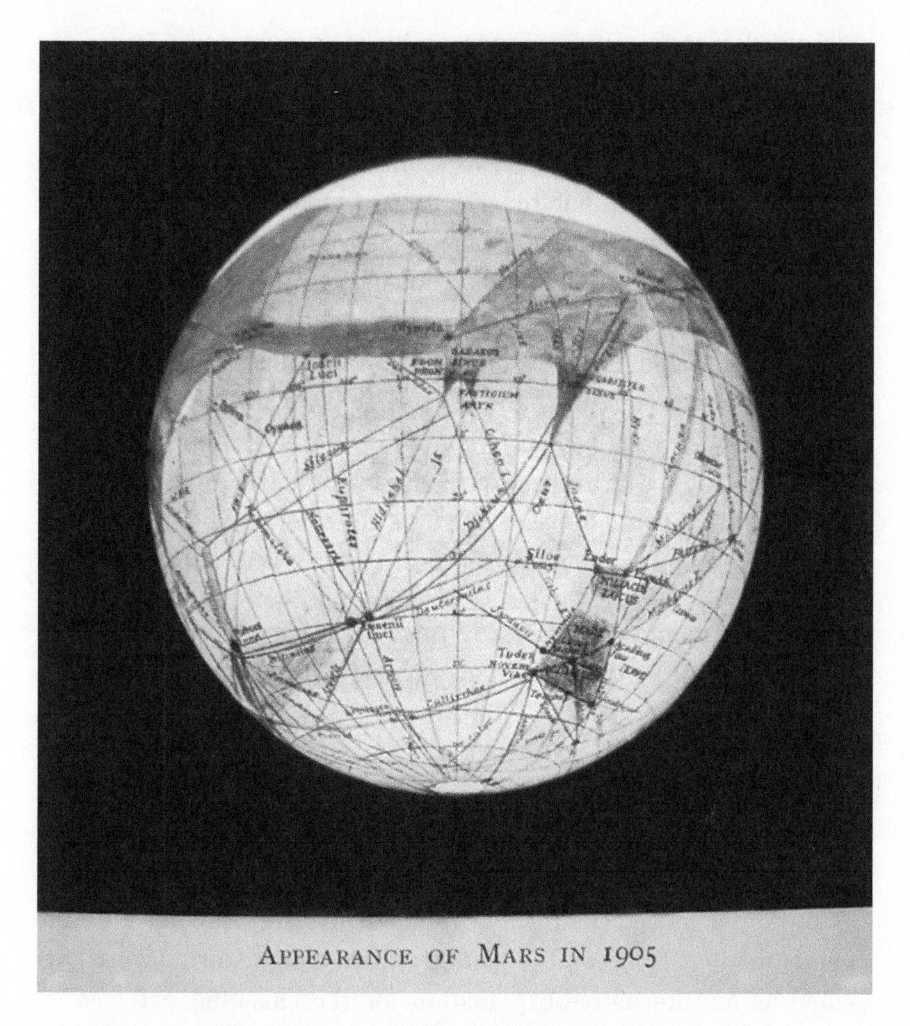

APPEARANCE OF MARS IN 1905

Fig. 7.3. Percival Lowell spent years sketching maps of Mars—and its canals—through his twenty-four-inch refracting telescope, perched high atop a freezing Arizona mountaintop. Sadly, it appears he may have in fact been charting the capillaries in his eye as reflected in the uncoated eyepiece. Well, everyone needs a hobby. Image from NASA.

Lowell was a contemporary of Wells, a passionate American amateur astronomer who wrote one smash book after another, advancing his theories on Martian civilizations. Each book was more colorful than the last, and by 1906, in his book *Mars and Its Canals*, he mused, "That Mars is inhabited by beings of some sort or other we may consider as certain as it is uncertain what these beings may be."[2] He further theorized that Martians had advanced transportation systems, technology well in advance of our own—including what we would now call television—and an organized and peaceful global government capable of irrigating the entire planet. For a few hopeful decades in the late nineteenth and early twentieth centuries, Mars seemed to be a place that might harbor our nearest neighbors in space, intelligent beings that could become welcome allies.

As astronomical tools improved after the turn of the twentieth century, telescopic observations revealed that the planet was generally drier and colder than Earth—this much could be gleaned using spectroscopy, the analysis of light reflected from the planet and its atmosphere, and other tools. But the fickle canals, observed by some astronomers and unseen by others, remained a puzzle. So was that "wave of darkening." Just how this might occur was a mystery, but the possibility that the canals were a Martian-designed irrigation system, feeding these seasonally blooming features, remained intriguing.

By 1964, most scientifically minded investigators had decided against intelligent life on the planet, and vast dust storms were suggested as an alternative explanation for the changing patterns on the distant world. But nobody could be certain. The romantics felt there was still a chance for a civilization of advanced beings, possibly human-like but adapted to their forbidding environment, to exist on Earth's neighboring planet.

That notion died when the signals from Mariner 4 were received at JPL, transforming this Victorian vision of Mars into a more modern one: the planet was a cold, cratered, and dead wasteland. The spacecraft had, within hours of its fast flyby, sent the first close-up views back to Earth—an event that changed everything.

By today's standards, the machine was very basic, but at the time Mariner 4 was state-of-the-art. The spacecraft was not an orbiter, but it would perform a speedy flyby of Mars, snapping twenty-one photos as it did so.

Fig. 7.4. An artist's impression of Mariner 4 approaching Mars. When this was drawn, Mars's true appearance was unknown, hence the smudgy "canals." Also, this early rendition of Mariner did not have the solar-wind steering petals affixed to the ends of the solar panels, but in the end, they weren't really necessary. Image from NASA.

Mariners 3 and 4 were of a completely new design—the new spacecraft was bigger than Mariner 2 and the Pioneers, more capable, and, perhaps most importantly, they carried a TV camera. This decision had not been a given—in fact, many scientists felt that a camera would be superfluous, opting instead for more radiation sensors, magnetometers, and other instruments that would return more detailed "squiggly line" data. But a few of the folks at Caltech, including a stubborn professor of physics named Robert Leighton, lobbied hard to include a TV camera. His reasoning included the notion that the tax-paying public should *see* what their money was being spent on. In the end, he got his camera.[3]

The improved design for Mariners 3 and 4 represented what some JPL engineers called the "branding" of the Mariner layout—this would end up being the basic structure they would use for a decade. John Casani, an engineer on the program at JPL, put it this way: "We wanted to standardize the layout for the Mariners, we likened in to a Volkswagen Beetle."[4] In those days, the diminutive German car covered just the basics of transportation, changing little as the years went by. "It was like building Volkswagens—they looked the same for years, with minor changes and improvements. Going from Mariner 3 and 4 to Mariners 6 and 7 was like a Volkswagen model change. Nothing major, just gradual improvement."

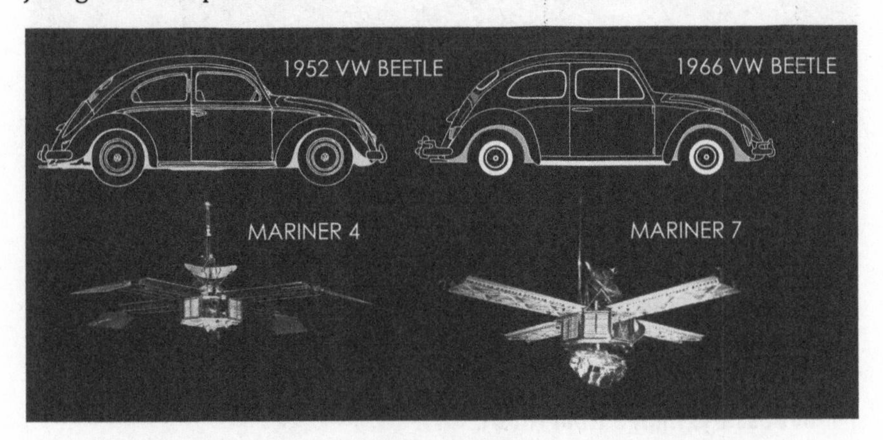

Fig. 7.5. John Casani's comparison is apt, even if a bit whimsical. Top left: 1952 Volkswagen Beetle, top right: the 1966 version. Bottom left, 1965's Mariner 4, bottom right, 1969's Mariner 7. See the similarities? Images from NASA and openclipart.com.

The basic Mariner design consisted of an octagonal frame made of magnesium, four feet across and about eighteen inches high. From that frame sprouted four solar panels—when deployed they stretched about twenty-two feet tip to tip. Atop this structure were two antennas, one high-gain dish, and a low-gain strut used when the dish lost contact with Earth. The TV camera was housed at the base of the spacecraft, and most of the instrumentation was inside the main framework, where it could be kept warm. The whole arrangement weighed about 908 pounds when fully fueled, right in line with the lifting capacity of contemporary US rockets.

Perhaps the most unique feature to people who have seen the early Mariners compared to later ones were the flaps at the end of each solar panel on Mariners 3 and 4. These were intended to be used for steering the probe in space without the use of maneuvering gas, of which it had a limited supply. Casani explained the rationale behind these rather cute appendages: "They were an experiment that didn't work. They were folded flat against the solar panels until the craft was in space, then when the panels folded down, the vanes deployed. ... They were supposed to swing out."[5] The idea was that the vanes could be adjusted to redistribute the pressure from the solar wind, the energy streaming out from the sun. In a vacuum like space, even the tiny pressure of these energetic solar particles might be enough to push the ship out of its proper orientation. By moving the vanes with little motors, the engineers thought they might be able to orient the spacecraft without using excess fuel to do so.

Casani continued: "We tested them on the ground, but only in the air, not in a vacuum chamber." And therein lay the problem: air is a damper, and the vanes were delicate things, and responded to the air in the test area by moving slower. "These were very light panels with thin Mylar sheets on them. In the vacuum of space there was no damping, and the inertia of the vanes' deployment carried them past their stops ... it might have stripped the gears, we weren't sure. In any case, the vanes went further than they were designed to." They unfolded and kept right on going until they folded over backward. "I'm pretty sure it would have worked, but not deployed that way!" In the end, they were not used on any of the succeeding Mariners.

Mariner 3 launched successfully—never a given in these early days—but the nose cone, also called a fairing, failed to separate properly, and the spacecraft died a prolonged death, enclosed in its aerodynamic cocoon with its solar panels folded up, as the batteries slowly ran out of juice.

The deficient fairing was redesigned in a few weeks, and Mariner 4 was launched in November 1964. Once in orbit, the fairing split in two and flew free as planned, and Mariner 4 was boosted toward Mars via a second burn of the Agena upper-stage rocket. While the launch and boost to Mars had been flawless—a relief after the issues that dogged Mariner 3—guidance soon became a problem. The spacecraft used two light sensors to track its position in space: one that tracked the sun and another intended to lock onto the bright star Canopus. The sun sensor functioned as planned, but the Canopus sensor had trouble staying locked onto that star. At first this seemed to be due to the sensor confusing the target star with others of similar brightness, but it was later realized that it was locking on to errant flecks of paint and dust that had bloomed around the spacecraft when it deployed from the rocket—the spacecraft had its own little constellation of light-reflecting crud lying in formation. New brightness-sensing parameters were sent up from JPL mission control, and navigation ceased to be a problem.

The spacecraft coasted through the darkness between Earth and Mars for seven months, with one course correction prior to its encounter with Mars. On July 14, the instruments were activated for the Mars flyby, and about seven hours before the closest approach the TV camera was switched on. This was done with great care—the primitive vacuum-tube video pickup could easily be burned out, and needed to be protected from any stray sunlight.

As Mariner 4 flew past Mars, images and data from the other instruments were saved to magnetic tape by an onboard recorder. As the probe passed the planet, it transmitted a continuous radio signal to Earth. This was the "Radio Occultation Experiment," in which signal strength was measured as it passed through the thin atmospheric haze surrounding Mars.[6] Precise measurements of the density of the Martian atmosphere were made for the first time during this experiment.

Then the spacecraft sped away from the planet, and the tape recorder played back data from the flyby. Each 200-by-200-pixel, low-resolution image took ten hours to downlink, and was sent twice to assure complete reception. It was the first use of digital imaging in planetary exploration, and while the images covered only about one percent of the planet, they were enough to seal Mars's fate forever.

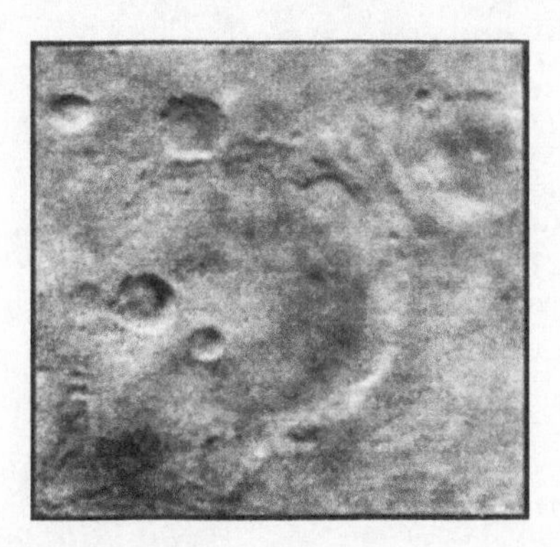

Fig. 7.6. In a single afternoon, Mars was transformed from a place of the imagination to a real world—and it looked an awful lot like the moon. The scientists were thrilled; the romantics among us, less so. Image from NASA/JPL-Caltech.

The pictures revealed a desolate, indescribably bleak, cratered landscape. Gone were the canals, the vegetation, and the Martian cities of the romantics. Mars was a dry, cold wasteland that looked more like the surface of the Moon than anything remotely resembling Earth. Its atmospheric density was only about one-hundredth that of our planet, about one-tenth of what was previously supposed. There was no liquid water to be seen, nor were there indicators of life of any kind.[7]

The destruction of a peopled Mars aside, the mission had been a smashing success. It paved the way for more Mariner missions to the red planet—Mariners 6, 7, and 9, the last of which would go into orbit there—and proved that NASA could perform complex operations at planetary distances. (Mariner 8 failed to reach orbit due to a rocket failure.)

Fig. 7.7. How it was done in the old days. A worker at JPL snips and clips images of Mars from Mariner 4. Scissors and glue have been replaced with computer imaging, but it can't be half as much "fun." Image from NASA/JPL-Caltech.

When Professor Leighton was later asked about the value of imagery in planetary exploration, he recalled a favorite moment. He had received a lot of fan mail from people excited about the mission, but one letter stood out. "I received a nice letter from a dairy farmer," he recalled. "He said, 'I'm not very close to your world, but I really appreciate what you are doing. Keep it going.'" Leighton smiled at the memory. "A letter from a milkman . . . I thought that was kind of nice."[8]

Mariner 4, with its grainy black-and-white images and rudimentary instrumentation, passed the red planet over fifty years ago. On that day, the Victorian view of the solar system vanished forever, and a new age of scientific understanding was born. Between the successes of Mariners 2 and 4, the road to the planets was wide open.

Chapter 8

FLASH FORWARD: MarCO

It was just a matter of time. CubeSats, developed by Stanford University and California State Polytechnic University, San Luis Obispo, in the late 1990s, had found their way into both commercial and NASA scientific applications. But for quite a while CubeSats had not found their way to other planets. NASA's new InSight Mars mission has changed all that.

On May 5, 2018, NASA launched Mars InSight, a stationary lander that will use seismometers to measure "Marsquakes" and attempt to understand more about the planet's interior structure. InSight will also deploy a self-burrowing, snake-like probe that will dig below the surface to measure the flow of heat from the planet's interior—another way to understand the interior structure of Mars.

But before any of that can occur, InSight has to land on Mars, and that's always tricky business. Any help is deeply appreciated, and two CubeSats riding along on the InSight mission are there to provide just that.

The CubeSat experiment is called MarCO, for Mars Cube One. They are the first CubeSats to fly into deep space. The intention is for the MarCO experiment to provide continuous status updates as InSight lands on Mars.

The CubeSat is a standard configuration used for small satellites, and is a cube about four inches to each side. Larger CubeSats are multiples of that—MarCO's design is a six-unit CubeSat, about the size of a briefcase. There are two of them stowed on InSight, MarCO-A and MarCO-B.

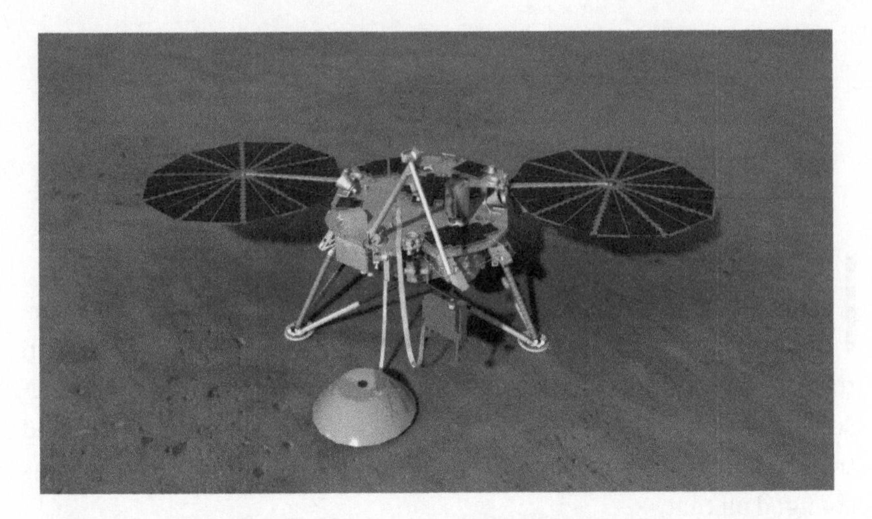

Fig. 8.1. Artist's impression of the Mars InSight lander seen after it has touched down on the Martian surface. To the front is a seismometer that has been placed on the ground via a robotic arm. It will be by far the most sensitive seismological instrument that's ever been used on Mars. Image from NASA/JPL-Caltech.

"MarCO-A and B are our first and second interplanetary CubeSats designed to monitor InSight for a short period around landing," said Jim Green, director of NASA's planetary science division at the agency's headquarters in Washington.[1] "However, these CubeSat missions are not needed for InSight's mission success. They are a demonstration of potential future capability. The MarCO pair will carry their own communications and navigation experiments as they fly independently to the Red Planet."

During InSight's challenging Entry, Descent, and Landing (EDL) operations, the lander will transmit information to NASA's Mars Reconnaissance Orbiter (MRO) flying overhead, and MRO will forward that data to Earth. But MRO cannot simultaneously receive information over one band while transmitting on another. Confirmation of a successful landing could be delayed by more than an hour due to MRO's role as a two-way radio link.

MarCO uses two frequencies and can both receive and transmit in real-time, allowing controllers quicker access to data from InSight.

To further complicate things, soon after InSight left Earth orbit, the two CubeSats separated from the Atlas V rocket that had launched the lander, in order to travel to Mars independently. To accomplish this, the MarCOs deployed two radio antennas and two solar panels. The MarCOs are being navigated to Mars separately from the InSight spacecraft to test their ability to travel and navigate through deep space—and they have been doing just fine, thank you. Each pint-sized spacecraft has eight small, cold-gas thrusters to allow them to make course adjustments on their way to Martian orbit.

Ultimately, if the MarCO demonstration mission succeeds, it could allow for a more robust communications relay option for use by future Mars missions during the critical few minutes of EDL, between Martian atmospheric entry and touchdown. MarCO will also pioneer the use of CubeSats outside Earth orbit, allowing the small, inexpensive spacecraft to be used for more ambitious planetary exploration missions.

Within the next few years, a number of projects using CubeSats to visit other planets are planned, some coming from universities and the private sector. Several groups and institutions are planning to explore the outer solar system with CubeSats.[2] The use of CubeSats—and even smaller nano-spacecraft—in deep space will usher in a new age of exploration and discovery in the solar system, and ultimately well beyond.

Chapter 9

LANDING ON LUNA INCOGNITA

Back to the mid-1960s we go! With the moon mapped as well as could be done by orbital cameras and crash-diving spacecraft, the next step in preparing for human landings was to send a robot to its surface. Besides wanting a better understanding of the makeup of lunar terrain in general, NASA needed to dispel any concerns that the moon might not be able to support the weight of the 16,000-pound Lunar Module of the Apollo program (that was its approximate Earth weight when landed on the moon with the descent module's fuel expended; its weight on the moon could be closer to 2,650 pounds in the reduced gravity).

Thomas Gold, a professor of astronomy at Cornell University, had been a NASA consultant since the beginning of the space agency in the late 1950s. He helped to design some of the cameras the Apollo astronauts carried to the moon and was considered of sound mind in all respects. But he had one major concern that alone merited robotic reconnaissance of the moon's surface—his hypothesis was that a thick layer of dust covered the moon, one so deep that a landing Lunar Module might sink out of sight forever, taking its crew with it, never to be seen or heard from again. While few in the planetary science community agreed with him, this was just one more reason to send the machines in first. So NASA did, with a project called Surveyor.

But Surveyor would not be first to land on the moon. Besides the deliberately crashed Luna probes, the Soviet Union sent more Luna missions to attempt a landing on the rocky satellite. It was successful on its twelfth attempt, Luna 9, in January 1966. After a three-day transit, the spacecraft was oriented with its landing rockets pointed toward the moon, and after inflating two bags to help cushion its impact, it fired its small rocket motors and set down on February 3. Luna 9 operated for eight hours, transmitting forty low-resolution images from the lunar surface. These photos showed a rough, rocky surface nearby, and it was clear that the spacecraft had not sunk into a deep layer of pillowy lunar dust, never to be seen again. Score one more for the USSR. But Luna 9 weighed only about 220 pounds, about 36 pounds on the moon. The Lunar Module was far, far heavier.

While the results from Luna 9 published in *Pravda*, the state-run Soviet newspaper, were helpful, the limited sharing between the Soviets and NASA meant that success with the Surveyor program would be critical. Studies that would culminate in the Surveyor program were started at about the same time as the Ranger program, in 1960. Within a year, the Surveyor program, like Ranger and Lunar Orbiter, was swept up in "space race fever," and with the announcement of the lunar landing program, NASA stepped up the pace on Surveyor as well.

Surveyor was originally envisioned as an orbiter and lander, but these programs were eventually split as the race for the moon continued to heat up. JPL was assigned oversight of Surveyor, with Hughes Aircraft selected as the contractor to build the robotic lander. Seven Surveyors were launched, with five ultimately landing on the moon, for a cost of about $470 million. This sounds like a lot in 1960s dollars, but, given the high stakes of sending humans to the moon, it was a pretty good deal. The Surveyors were, after all, the reconnaissance flights for a number of specific Apollo landing targets, and more generally of interesting types of lunar terrain.

The Surveyor spacecraft was a spindly looking affair with three landing legs, a tubular frame, and two flappy solar panels at the top. Later Surveyors added a scissors lift–like robotic arm to test the nature of lunar soil. By the time it had ejected the retrorocket pod and primary radar unit before landing, the spacecraft weighed about 650 pounds (or about 107 "lunar pounds") upon touchdown. Since the first Apollo landings were envisioned to be as early as 1967 or 1968 at the time, the robots had to get there early enough to inform the decision-makers working on the manned program.

In a welcome shift from Ranger, the very first Surveyor successfully landed on the moon. It was launched on an Atlas rocket, the repurposed nuclear missile that had also launched John Glenn into orbit in 1962. Like the Mars rovers that would come decades later, the Surveyors did not go into lunar orbit before landing, but were sent in a trajectory that aimed right at their landing spot on the moon.

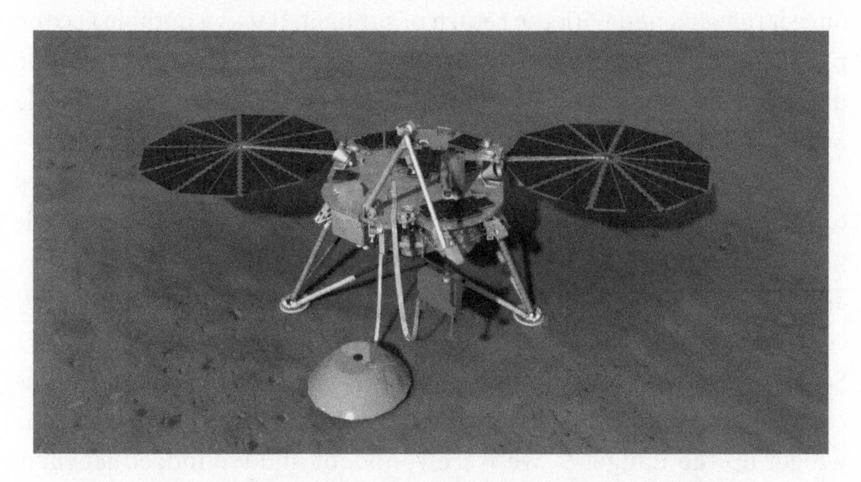

Fig. 9.1. A Surveyor mockup being tested on Earth. It would have looked similar once it landed on the moon, sans the dead grass in the background. Image from NASA/JPL-Caltech.

About sixty-three hours later, and forty miles above the lunar surface, still traveling at suicidal trans-lunar speeds, a solid rocket fired for forty seconds to slow the spacecraft. Then, over the next two and a half minutes, smaller liquid thrusters were fired to continue slowing and steering the craft to touchdown. By the time it was about fifteen feet above the moon, Surveyor 1 had slowed to nearly zero velocity, and the rockets were shut off, allowing the probe to slowly fall to the surface under the moon's one-sixth Earth gravity. The point of this was not to tempt fate, but rather to avoid disturbing the lunar surface with the rocket blast.

Surveyor 1 and Surveyor 3 both landed on Oceanus Procellarum, the Sea of Storms, which would, about two years later, be the landing site of Apollo 12. Surveyor 1 was the first soft landing of a US spacecraft on another world, and came about four months after the USSR's Luna 9.

In its short life, Surveyor 1 transmitted 11,240 images back to JPL. It lasted a few weeks on the lunar surface, the first spacecraft to survive the chilly two-week lunar night, and sent the last 800 images once it reawakened with the return of sunlight. It was a notable accomplishment on a world where temperatures at local noon can reach 240 degrees Fahrenheit, and plunge to 290 degrees below zero in darkness. The mission was declared complete on January 7, 1967.

There had been some concern about the news media with regard to the Surveyor landings. This is not surprising, given JPL's struggle with the Rangers. But NASA had been broadly accepting of the press, and JPL would be sorely out of step if it banned the media from live coverage of the event—though they were entirely capable of doing so. Said Bill Pickering, then director of JPL, "We had started out with some concern about letting the networks handle this live, but by the time we got up the Ranger 9, we were confident and we indeed let them [in]."[1] But Surveyor renewed the concerns. "When it came time for the first Surveyor, which was to actually land gently on the moon and then look around and start taking pictures, we were much more nervous. So we were not at all sure we wanted the networks to really do that live. But we finally ended up by agreeing to let them do it, and we kept our fingers crossed and hoped it was going to be all right." Pickering laughed as he recounted this in a 1978 interview. "But the thing that startled me was that about a half an hour before it was due to land, one of the network people said, 'Oh, by the way, we're live all over the world,' which really sort of shook me. Fortunately, it worked."

Bravo for Surveyor 1. In contrast, the second Surveyor crashed near Copernicus crater when its landing rockets failed to ignite. But the third flight, Surveyor 3, landed not far from the Surveyor 1 site in April 1967. Surveyor 3 was notable for a few reasons. It was the second successful landing in the program, and it included a robotic arm with which it could nudge and scoop lunar soil while the TV cameras watched, giving scientists on Earth a limited understanding of the nature of the lunar surface and its possible bearing strength. More notably, Apollo 12 landed about 1,200 feet away from Surveyor 3 in November 1969, a remarkable feat of navigation so early in the Apollo program. During their moonwalks, the crew walked over to the Surveyor and snipped off its TV camera with a pair of bolt cutters brought along for that purpose—it was the first and only time that humans visited one of their own robotic probes on another world. Upon return to Earth, the camera was analyzed to see what changes had occurred from a couple of years on the moon—they were minor, though it was noted that some bacteria, which had been resident on the camera at launch in 1967, appeared to have survived its time on the moon.[2] Surveyor 3 operated for about two weeks.

Surveyor 4 failed in flight, but Surveyor 5 landed in Mare Tranquillitatis, the Sea of Tranquility, where Apollo 11 would land less than two years later. It returned over 19,000 images and carried the first true scientific instrument to be used on the lunar surface, an alpha-backscatter device that had the ability to do basic chemical analysis of the nearby terrain. This Surveyor also survived a lunar night, operating for another twenty-two hours after the return of the scorching sun.

Surveyors 6 and 7 were also successful, each with unique results. Surveyor 6, which landed in November 1967, conducted experiments similar to its predecessor, but also moved itself to a new location, albeit just a few feet away. After conducting its first set of investigations at its initial landing site, its descent rockets were fired again, allowing it to hover and land about eight feet distant. It was not much of a move, but at the time this represented an unprecedented amount of control over a landed robotic craft. The robot was also able to then send back pictures of the first landing site, showing the disturbance to the soil from the secondary rocket blast.

Finally, in January 1968, Surveyor 7 landed in rougher terrain than any of its kin, near the rim of the huge crater Tycho. About 21,000 images were returned in the first day alone. The spacecraft was also able to observe the Earth and stars as seen from the moon—another first. The onboard cameras later spotted the flash of a laser beam shined from Earth, a prelude to a later experiment carried by Apollo—a reflective mirror box—that would allow astronomers to pinpoint the distance from the Earth to the moon within inches.

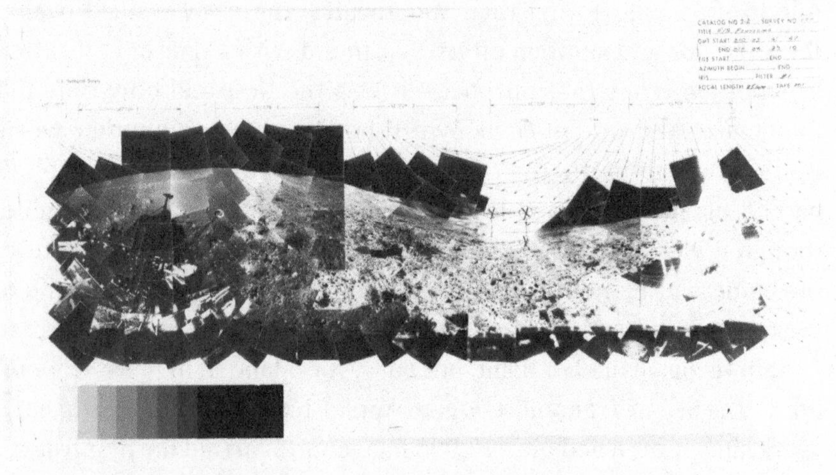

Fig. 9.2. An image mosaic sent back by Surveyor 7. The assembly of the multiple still photos is obvious—NASA ordered Scotch tape and glue in bulk in those days. Image from NASA/JPL-Caltech.

All told, the Surveyors operated for an accumulated seventeen months on the moon and returned 87,000 images for examination. While modest compared to the results gleaned from modern Mars rovers or the Voyager program, some of which have operated for a decade or more, for the time the Surveyors were a notable achievement in command and control, as well as hardware capability.

The stage was set for a manned lunar landing in 1969, but that was something for other departments of NASA to set their sights on. JPL had an ongoing relationship with Mars, and that program was ramping up to attain its own spectacular heights.

Chapter 10

RUSSIAN ROVERS

Other than the Luna and Surveyor landings, most of the intensive exploration conducted on the surface of the moon was done by the Apollo astronauts between 1969 and 1972. The astronauts explored the area of a medium-sized town across their six successful landings. At about the same time, from 1970–1973, the Soviet Union sent a series of robotic rovers to various points on the moon. These were the Lunokhods (Russian for "moon walkers"), and looked like large washtubs on eight wheels.

The Lunokhod program was initiated as a backup to the struggling Soviet effort to land humans on the moon before the Apollo program could do so—the core of the space race decade. The Soviet lunar rovers were originally intended to reconnoiter the first landing sites of the cosmonauts a few weeks before their arrival on the moon, but the Soviet human landing program failed spectacularly when their huge moon rocket, the N1, failed during flight testing (not just once, but four times). With the human program in disarray, the Lunokhods persevered alone.

The first Lunokhod was launched in February 1969, but the rocket exploded, taking the rover with it. A second attempt was made in November 1970 and was a success, becoming Lunokhod 1—this designation was in line with the USSR's habit of renumbering missions to reflect successful outcomes. So rather than naming the first Lunokhod to land successfully on the moon Lunokhod 2, it was named Lunokhod 1.

The 1,800-pound machine was the first successful robotic rover to travel to another world, and set the standard by which other rovers would be measured. It carried an array of TV cameras and extendable appendages to conduct experiments with lunar soil. Lunokhod was powered by batteries that drew power from a solar panel mounted to the inside of a lid that covered the "washtub"—an ingenious design that kept the solar panel clean during landing and the interior warm during the long lunar night. The lid was opened for operations during the lunar day, then closed for the two-week night when temperatures plummeted. A small clump of heat-generating radioactive polonium was housed inside the rover, and with the lid closed this kept the internals warm. The Lunokhods were large machines, measuring almost eight feet in length and nearly five feet high.

Fig. 10.1. The Soviet Lunokhod 1. They were not destined to win any styling awards, but the lunar rovers were effective. Image from NASA/GSFC/Arizona State University.

Lunokhods were instrumented well for the era. Each carried an X-ray fluorescence spectrometer that could analyze surface samples, and an X-ray telescope to attempt astronomy from the lunar surface without the interference of Earth's atmosphere. A radiation detector was included to measure cosmic radiation, as were various devices to measure the hardness of the surface. The rovers also had a retroreflector device to allow the measurement of the distance between the Earth and the moon, similar to those deployed by the Apollo missions.

Fig. 10.2. Lunokhod 1's lander after the rover had driven off, turned around, and snapped a photo. Note the tracks to the right of the ramps—they are still there, and will be for millions of years. Image from NASA.

Two Lunokhods were successfully landed and operated on the moon. Once each spacecraft reached the vicinity of the moon, it went into lunar orbit before proceeding to land. This allowed Soviet scientists to optimize their plans before committing to the final landing site.

Lunokhod 1 operated for 322 days, covering about 6.5 miles of terrain during that time. It returned 22,000 images and lots of scientific data about the lunar surface, with twenty-five sets of soil analysis. While it was slow, like all robotic rovers, it was able to loiter at areas of interest far longer than the Apollo astronauts could stay at one place, allowing for a leisurely look at the specific target.

Lunokhod 2 reached the moon in January 1973. While it only operated for about four months, the rover covered a distance of almost

26 miles in that time, a record for robotic traverses that stood until NASA's Opportunity Mars rover surpassed it by a slight margin—as of 2018, Opportunity's total distance traveled stands at 28.06 miles, but that's taken fourteen years. Lunokhod 2 also drove across more challenging terrain than Lunokhod 1, covering patches of moon sand and exploring lunar rilles, fissures caused by volcanic activity. It returned over 80,000 images and conducted a number of important experiments on lunar soil and rock.

The Lunokhods came back into the public eye over a decade later, when the Chernobyl nuclear power plant exploded in 1986. Existing robotic investigation machines were unable to enter the rubble produced by the explosion, and members of the Lunokhod engineering team were reassembled to design a robotic rover that could safely enter the plant—there was initially too much radiation present for people to work within it for more than a minute or two. While electronics were a concern, as radiation will wreak havoc with them, the Lunokhod-sourced units were already designed to be hardened against radiation due to the conditions found on the moon, and performed well in the high-radiation environment.

In a final twist, the ownership rights to Lunokhod 2 were put up for auction in 1993 through the Sotheby's auction house. The successful bidder was video game entrepreneur, space enthusiast, and space tourist Richard Garriott. He paid $68,500 to own . . . well, nothing . . . except for the hardware on the distant moon. But, as he put it, "I purchased Lunokhod 21 [the technical name for the lander] from the Russians. I am now the world's only private owner of an object on a foreign celestial body. Though there are international treaties that say no government shall lay claim to geography off planet earth, I am not a government. Summarily, I claim the moon in the name of Lord British!"[1] "Lord British" is a fictional creation in Garriott's video game *Ultima*. And why not? Garriott may find his claim contested when private entities begin laying claim to mining rights on the moon in the next decade or so, but claiming the moon in the name of a fictional character in a fictional world should not upset anyone . . . for now.

Chapter 11

FLASH FORWARD: PROSPECTING THE MOON

There's a lot of renewed activity around the moon today, most of it robotic in nature. While much of this centers on the Trump administration's new directives for NASA to prepare for a return of humans to the moon, or at least its orbit, with a lunar orbiting station called the Gateway,[1] there is also interest in lunar activity within the private sector.

This interest has been slowly building for a couple of decades, but was kick-started by a competition ultimately called the Google Lunar XPRIZE (GLXP). This competition was a part of the larger XPRIZE competitions, which started with the Ansari XPRIZE. This original XPRIZE was announced in 1996 with a cash award of $10 million for the first nongovernmental organization to fly a reusable crewed spacecraft into space (which officially starts at an altitude of 62 miles) twice within two weeks. Aerospace entrepreneur Burt Rutan won this in 2004 with his SpaceShipOne, financed by billionaire Paul Allen.

With this successful project behind them, the XPRIZE Foundation partnered with Google to offer the GLXP with underwriting from Google. This time the purse totaled $30 million in various awards, and the challenge was for a privately funded nongovernmental team to fly a robotic machine to the moon, travel 500 meters (about 546 yards), and transmit hi-resolution imagery back to Earth. The GLXP was announced in 2007, with this kickoff statement by prize founder Peter Diamandis, a new space entrepreneur: "It has been many decades since we explored the moon from the lunar surface, and it could be another six-to-eight years before any government returns. Even then, it will be at a large expense, and probably with little public involvement."[2]

The GLXP was extended and extended again, as the teams struggled to reach their goals, until it finally expired for the last time in March 2018 . . . with no winner. Thirty-three teams entered, and many persevered for years. The field narrowed as the complexities of the project—mostly with regard to launching into a trans-lunar trajectory—became apparent, but in the end, the GLXP was ahead of its time. By the time of its cancellation, five teams were still laboring away. Finalists included Moon Express from the United States, Space IL of Israel, TeamIndus out of India, Team Hakuto from Japan, and Synergy Moon, which a represented an international partnership of numerous countries. Moon Express had been at it since 2010, and had a lot of money behind them courtesy of founding billionaire Naveen Jain. The team built a number of scale and full-sized prototypes, and along the way even acquired another GLXP competitor and their technology. They also procured a NASA contract for data return from their spacecraft that was worth a substantial sum. Their facility expanded from their Mountain View, California, headquarters to two leased spaces at Cape Canaveral, and they seemed to be poised to win the competition with comparative ease. In 2015 they had signed agreements with small rocket startup Rocket Lab for three launches, and it appeared that they would handily beat the deadline. They even negotiated an agreement with the Federal Aviation Administration (FAA) for a commercial/private mission beyond Earth orbit, another first.

Sadly, for a variety of reasons, Rocket Lab fell behind schedule and, while they were able to launch orbital tests before the deadline, even delivering commercial payloads into space, the Moon Express launch did not occur. The GLXP was closed quietly with this announcement by its founders: "After close consultation with our five finalist Google Lunar X Prize teams over the past several months, we have concluded that no team will make a launch attempt to reach the moon by the March 31, 2018, deadline," they said. "This literal 'moonshot' is hard, and while we did expect a winner by now, due to the difficulties of fundraising, technical and regulatory challenges, the grand prize of the $30M Google Lunar X Prize will go unclaimed."[3]

But this has done little to deter the private sector in their plans to harvest resources from the moon. Moon Express is just one player in this area, with plans that include lunar-based astronomy, other experiments carried for science customers, and ultimately lunar resource extraction. They plan to begin operations on the moon by late 2019. Some of the other competitors continue to move forward as well. Team Hakuto recently signed a tentative agreement for SpaceX to launch their spacecraft in 2020 or 2021, for example.[4]

There are many other companies and organizations exploring the development of the moon beyond the finalists of the GLXP. Prominent among them is Blue Origin, the innovative private aerospace company created by Amazon founder Jeff Bezos. Blue Origin has been in existence since 2000, and has developed and flown its reusable New Shepard rocket eight times, with great success. In 2017, the company announced that it was exploring a new project it called Blue Moon, a planned robotic cargo lander that would fly to the moon aboard the company's much larger rocket, the New Glenn. The Blue Moon lander would be capable of delivering 10,000 pounds of cargo to the moon, and could additionally return cargo—possibly derived from lunar resources—to Earth orbit. Current plans call for the lunar lander to be flown in 2020.[5]

There is a lot more activity in this sector, and it's developing quickly. Investment in private space ventures is at an all-time high, and the moon is a prime target for resource development and commercial activity. With no pun intended, watch this space.

Chapter 12

THE CENTER OF THE UNIVERSE: PART 2

As Mars filled the video screens at JPL on August 5, 2012, the assembled media congregated in JPL's Von Karman Auditorium for the final act of the landing of Curiosity. I picked a spot toward the front of the auditorium to get a ringside view of the descent of the spacecraft to the Martian surface. There was a lot riding on this—the project had blossomed to $2.5 billion over the years, and unlike many of JPL's previous high-risk missions there was only one chance at success—Curiosity flew alone. The Mars Exploration Rover (MER) mission of 2004 flew twin landers, Spirit and Opportunity, to expand their chances of success and to explore two differing regions of the planet. But Curiosity was an expensive beast, and at almost two thousand pounds, a monster of a machine to deliver safely to Mars.

The MER rovers were a comparatively diminutive four hundred pounds each, and had been designed as a more capable follow-on to the experimental 1997 Mars Pathfinder mission. In each of those cases, the machines had been delivered to Mars via the use of shock-absorbing airbags after a long and exhaustive research effort led by Rob Manning, the chief engineer for all of JPL's Mars rovers.

A week or so after the landing, I would return to JPL for a post-landing conference. There would be a number of these, but their frequency decreased in the weeks after touchdown, not because of any lack of generosity on JPL's part, but because the news cycle had moved on. Soon it will just make more sense to hold such press events as online events, because the media that will pay attention beyond a sound bite become more specialized and smaller in number. On that day, about a week after landing, JPL's mall area was still abuzz, with the press there in force to await new discoveries and ogle the spacecraft on display, which include a full-sized Curiosity mockup. Nearby, and somewhat incongruously, a jazz band started playing. This was not just any jazz band, however; it was a JPL jazz band, and composed of some of the smartest people you're likely to meet at either a smoky back-room club or a NASA facility. A good deal of the people who work at JPL de-stress their intense minds through various hobbies . . . you know, relaxing stuff, such as rock climbing, aerobatic flying, advanced karate training, or even playing jazz. In that afternoon's ensemble were the Manning brothers, a rare set of identical twins who both work at JPL. On the saxophone was Chuck Manning, who is a process engineer at JPL's Microdevices Laboratory, where many of the tiny chips and slivers that make up the world's most advanced spectrometers and sensors are manufactured. Some are actually sculpted at an atomic scale, with electron beams acting as the chisel. Chuck oversees the sophisticated machines that do this work, along with a few dozen other areas of endeavor. Oh, and he's also a world-class jazz musician, having played with some of the biggest names in the business. A few years from that time, I'd have the opportunity to work with him on some technical outreach projects, an engaging and enjoyable process.

On trumpet was his Caltech-educated brother Rob, who was the chief engineer on Curiosity. He's been in that post for every Mars rover built, after cutting his teeth in projects like Cassini. "My first decade at JPL revolved around spacecraft computers and advanced computer architectures," he says.[1] "It was fascinating work, but technologically a long way from computers as intelligent as HAL from *2001: A Space Odyssey*. I went on to be the lead engineer for the computers used on the Cassini spacecraft. Later I found that designing fault protection would be a lot of fun, and spent two interesting years thinking about how to make Cassini diagnose failures and repair itself during its long, lonely voyage to Saturn."

Fig. 12.1. Artist's impression of a MER rover as it prepares to drive off its lander after bouncing to a landing on Mars in 2004. In this image the deflated airbags are neatly retracted; in practice, they looked more like the tangle of a teenager's unmade bedsheets. Image from NASA/JPL-Caltech.

By the time I was hired as a contractor at JPL, Rob Manning would be ascending to nosebleed territory in Building 180, as chief engineer not of a project but for all of JPL. And he's still one hell of a nice guy.

As I listened to the tunes, I noted that this place was annoyingly full of overachievers—some of the best technical and engineering minds in the world. I was not one of them, but at least I was there to hang out with them for the day. After Rob completed his set, I asked him if we could chat about the landing phase of Curiosity. "Sure," he said—he was still rightfully amped about the mission. We started by discussing the MER rovers.

"The original idea was to have vented airbags like the ones in cars that would deflate right away, so all this energy is dissipated and is more like a sandbag. But when we realized we couldn't get the airbag vents to work, [fellow engineers] Brian Muirhead, Tom Rivellini, and I all agreed that we would be bouncing." They had tried a dozen ways to get the airbags to deflate immediately, but the results were unsatisfactory. So they would simply let the enclosed rover bounce across the

Martian surface until the energy of landing was diminished. "We knew we were going to bounce anyway . . . so let's just bounce multiple times and make sure the system can handle it."[2]

They tested their airbag system exhaustively until they were sure they had it right, and used it first on the Pathfinder spacecraft. When Pathfinder arrived at Mars, it bounced merrily along to a final resting place, from which the mission proceeded successfully. Then they repeated the method with the MER rovers, with equal success. This worked well for the four-hundred-pound rovers, but would be a bust for Curiosity, which was far too heavy and delicate to use the airbag landing method. Manning and his team considered the alternatives, including the use of a traditional lander platform that the giant rover could drive off of.

"If we used airbags, the rover would have to climb off the lander and get past the airbags," Manning said, and that was a possible trap. Climbing over airbags might cause the rover to get caught up like a person swimming into a fishnet. "Additionally," he added, "if you land on a propulsive lander you're sitting way up high and it's hard to land when all your weight is up high. You also need ramps or some other way of descending to the surface, which is much more complicated." And complication equals an enhanced possibility for things to go wrong.

They pondered alternative designs—some truly unusual: "This was thinking way out-of-the-box . . . we were literally going through all possible ways trying to imagine every possible configuration, whether it made sense or not. We did this until we bumped into one that seemed crazy, but that actually made sense." If the spacecraft was too tippy and complex with the rover on top of the lander, why not invert the combination?

"Ultimately, the egress problem was solved by taking the rocket descent stage—from which a rover would traditionally have to disembark—and putting it above the rover," he said.

Curiosity, then, was essentially an inverted version of other rover designs, like those used for MER, Pathfinder, and even the Soviet Lunokhods. Instead of a rover riding to the surface on a lander platform, from which it would drive down to the Martian surface, the designers placed the rocket pack on top of the rover. This eliminated a lot of the complexity, made it more stable, and allowed the rover to land on its wheels, removing the need for a separate set of landing gear. Many birds were killed with one technological stone.

But the design introduced other complexities. Curiosity would be hurtling into Mars's atmosphere at about 13,000 miles per hour, and it had to somehow be slowed to a walking pace before it reached its selected landing zone at Gale Crater. Gale had been selected for a combination of attributes, including the low altitude of the crater floor—where the atmosphere is slightly denser, and which gives the spacecraft more time to slow down—and the promise of great geological diversity and the likelihood that there might once have been water there for millions of years.

So the final design of the landing—the EDL (Entry, Descent, and Landing)—would go like this:

As it neared Mars, the aeroshell containing Curiosity—which looked like a flattened Apollo capsule—would eject the cruise stage, the ring that sat atop it and contained the avionics and maneuvering equipment that had gotten it from Earth orbit to Mars. Curiosity would then be a neatly streamlined package.

Upon reaching Mars, Curiosity would slide into the atmosphere, and ultimately glide through the air with its elongated heat shield, which was designed to create aerodynamic lift. In this phase it would actually be flying, actively navigating to the landing zone. Guided by its internal computer, the spacecraft would use a combination of maneuvering jets and periodically ejected tungsten weights to change its orientation. The maximum heating experienced by the heat shield would be almost four thousand degrees Fahrenheit.

After Curiosity slowed to about Mach 2, the supersonic parachute would be deployed. Designing and testing this was a nightmare in itself, as simulating a descent through the thin Martian air at supersonic velocities is extremely tricky on Earth. The engineering team tried high-speed wind tunnels and high-altitude test drops, but in the end there was a certain amount of faith involved. Computer simulations were simply unable to accurately model the behavior of cloth inflating suddenly at high speeds—it behaves as much as like a wild animal as the way a parachute canopy would on Earth.

With the parachute open and doing its job, the heat shield would be dropped. Then, after a few more minutes of descent, when the spacecraft had slowed to about 270 miles per hour, Curiosity would drop free from the parachute and aeroshell.

The landing rocket pack would be perched atop the rover, looking like an oversized backpack with small rockets arrayed around the perimeter. Those rockets would fire, slowing the rover to a hover at about sixty-five feet above the floor of Gale Crater.

Then the real weirdness would happen.

As the machine hovered, the rover would detach from the rocket pack and be lowered by four tethers, one attached to each of its extremities. It would rappel down to the surface, and when the wheels made contact, a signal would go to the rocket pack, telling it to detach and continue firing long enough to fly a few miles away and crash.

Fig. 12.2. Immediately upon touching the surface of Mars, Curiosity sends a signal to the sky crane rocket pack, which detaches its tethers and rockets off to crash a couple of miles away. At least, that was the plan. Image from NASA/JPL-Caltech.

At that point, Curiosity would be on Mars, safe and sound, with very little contamination from the rockets—remember that they remained sixty-five feet up in the air as they were firing—and would be able to begin its exploration of the pristine Martian surface right where it was. No lander to drive off of, no bulky landing bags to drive over, just turn on the cameras and go.

Back on landing day, that was what was supposed to happen. What would actually occur when this seemingly Rube Goldberg arrangement was tried for the first time on another world was anyone's guess. I had pigeonholed Manning about a week before the landing, and asked about the likelihood of success.

"How do you know when you're done testing? This is a real problem, and it drives management crazy," he said with a chuckle. When talking to his managers, he will often say, "We haven't tested this, and we haven't tested that, but basically when you run out of reasons why it *should* fail, that's when you can stop testing." He added, "Many things can still go wrong, but I've tested everything I can, I've gone through and looked at everything, and eventually, the items you can find on your list of things to do goes to zero." I'm a bit slack-jawed—I'd always assumed that the process was a bit more *certain*. He clearly enjoys my confusion. "Of course there are still plenty of things that could go wrong," he says, "That parachute can still get hung up, for example, but we've done everything possible to assure success." I asked if he was worried, and he just smiled.[3]

Back at JPL, I moved back inside the auditorium now that darkness had descended on the mall. The reporters, TV people, and bloggers were more attentive now, as there were only a few hours until atmospheric entry and the landing sequence. Journalists who had not paid a lot of attention to how the mission worked were looking at the press packages—many with furrowed brows. Most all of them had seen the "7 Minutes of Terror" video by now, but seeing the sequence on paper revealed more complexity. There was a lot to digest.

The moment of truth was just three hours away, and, on the screens dominating the end of the auditorium, the time to landing was counting down quickly.

Chapter 13

VOYAGERS ON MARS?

In 1972, with the moon landings behind them, NASA turned loose increased funding for the most ambitious Mars mission ever with the twin Viking orbiters and landers, scheduled for a 1976 landing. But the massive program had its origins back at the beginning of the space race, and by another name . . . one that will be familiar. The first plan for a robotic Mars landing was originally called Voyager.

So far, Curiosity has been the largest NASA robotic probe to land on Mars. And few are likely to be larger—the difficulties of landing heavy payloads on Mars is one factor, and the continuing miniaturization of spacecraft is another. What used to be large and heavy is now becoming smaller and lighter, thanks to rapid advances in the miniaturization of circuits and investigative machinery. Mars 2020, Curiosity's follow-on rover, may be the last of the large, heavy robots to be delivered to Mars for some time. At 2,314 pounds, it's pushing the limits of safe delivery, and at the time of this writing, the parachutes are still being tested.

A possible exception to the above may come from the private sector—from time to time, Elon Musk of SpaceX fame had made noises about sending one of his Dragon 2 space capsules to Mars in an unmanned configuration. He has more recently switched gears, concentrating instead on a Mars journey with his "Big Falcon Rocket," a rocket and spacecraft that dwarfs the Apollo era's Saturn V. He may attempt a landing on the planet, either robotically or with a crew, as early as 2024. As of now, however, there are no firm plans on the table. As Musk likes to say, the plans are "aspirational."

But travel back in time a bit, and you can find plans for truly massive planetary exploration machines. One of them was called Voyager Mars, and it was headed not for the outer solar system, but to the red planet.

In 1960, JPL began to study how it might send a robot to Mars. As we've learned, not a lot was known about the planet at the time—only what could be gleaned from telescopes and associated instrumentation. The first robotic flyby was still five years in the future, so the mission designers would have to make a lot of assumptions. But the space age was in full flower, and assume they did—with some rather dramatic, but unavoidable, inaccuracies.

Remember that at this time NASA had only existed for two years. JPL was older, having been founded in the 1930s to investigate the adding of jet engines to airplanes to help them take off on shorter runways—hence the "Jet Propulsion" part of the name. In 1936, some Caltech grad students began experimenting with rockets in Pasadena's Arroyo Seco, a miles-long gulch that runs along the western border of the city. By 1941, this research had expanded to experimentation with Jet-Assisted Takeoff (JATO) for airplanes, and by 1943 the US Army had engaged with Caltech to form the Jet Propulsion Laboratory.

The experiments moved from place to place—rockets and experimental jet engines were dangerous, and mishaps not uncommon. At one point, an explosive failure on the Caltech campus prompted the relocation of these experimental activities to the foothills at the head end of the Arroyo, where JPL now resides. For two decades JPL had worked under the Army, until NASA was formed in 1958 and the Caltech-managed Lab was then transferred to NASA oversight. With the flight of Sputnik just a few months earlier, the space race was on, and JPL would have an immediate role—formulating a crash program to build Explorer 1 (after the failure of the Vanguard launch attempts) to send America's first satellite into orbit.

Given the quick success of that undertaking, it was not surprising that the Lab was charged with most of NASA's robotic planetary exploration. As we've seen, the moon was of early interest, and the nearest of the planets, Venus and Mars, soon followed.

Fig. 13.1. In the lobby of JPL's Von Karman Auditorium stands a model of Explorer 1, America's first satellite. It was built in about three months' time, after the Navy's Vanguard rocket failed in 1957. Image from NASA/JPL-Caltech.

The first Voyager proposal suggested orbiters and landers for both Mars and Venus, machines that would hopefully provide stable, long-lived platforms for an in-depth study of both worlds. The early flyby reconnaissance of those planets by the Mariners would inform the design of the far more ambitious Voyagers. But the realities of the space race interfered with this plan—most of NASA's dollars were being spent in an attempt to catch up with the Soviet Union in human spaceflight, and, at the same time, problems with the rockets that were needed to launch even small probes were delaying the Mariners. The boosters were all too often as happy to veer off course and come careening back, head first, toward the launch complex as they were to go up, where they belonged.

By 1962, with Kennedy's lunar challenge to the Soviets well seated in the public mind, Congress looked at the nation's space budget and decided to spend the vast bulk of it on manned lunar landing efforts.

JPL had to narrow their goals and lower their sights. After performing a bit of planetary exploration triage, the Lab decided that, while Venus would be attempted by the first launches of the Mariner series of spacecraft, when it came to Mars—which appeared much more Earth-like through the telescopes of the day—JPL would do something different. Evolved Mariners would still perform the first flybys, but the immediate follow-on would be much larger and more ambitious. This would become the Voyager Mars project.

The available data about Mars hinted at a world broadly similar to Earth in many ways. It had a solid surface, like the other terrestrial planets (which include Mercury, Venus, and Earth). Mars also had an atmosphere, but not one as dense and dangerously hot as that of Venus. While it was clear that there was not much oxygen in Martian air, telescopic spectroscope readings indicated that it had abundant carbon dioxide, which at the time was thought (erroneously) to be possibly about the same density as that experienced atop Mount Everest. It was cold, but apparently not so much that some form of life would be impossible. As previously noted, while few scientific minds still thought there was any real chance for the existence of the advanced, intelligent civilization envisioned by Percival Lowell, many thought there might be plants and low-level organisms. Armed with these loose facts, mission planners decided that Mars was the worthiest target among the nearby planets for investigation, and the first mission proposed to land there was a doozy.

While the construction of spacecraft such as Surveyor had been farmed out to the aerospace industry, the Voyager Mars project would be an in-house undertaking by JPL, just as the Rangers and Mariners were. The plan was to build the Voyager orbiters based on lessons learned from the Mariners, and the landers would incorporate experience from the Surveyor missions. Within the confines of JPL, this was all deemed as manageable and doable.

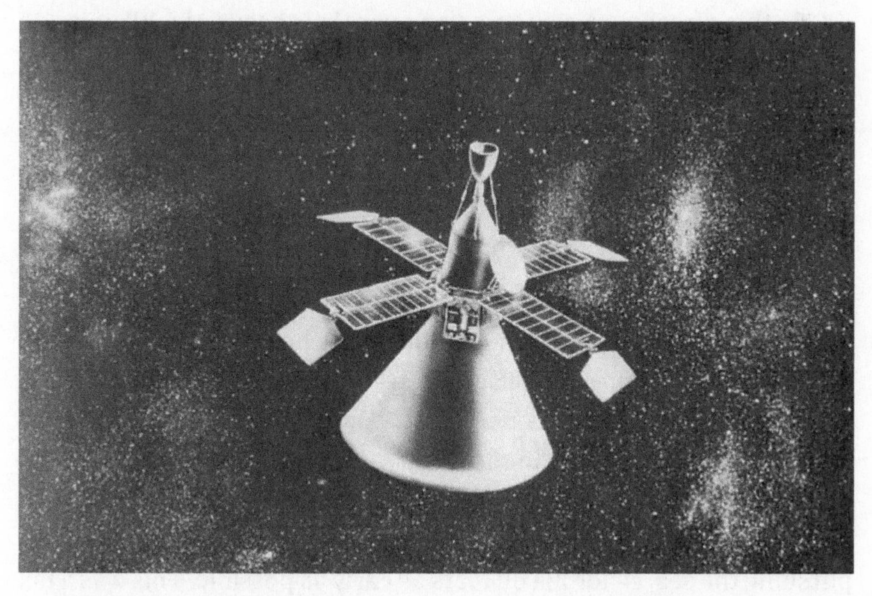

Fig. 13.2. An iteration of the ever-expanding Voyager Mars spacecraft. Note the Apollo capsule–style lower half. Image from NASA/JPL-Caltech.

NASA, on the other hand, was developing a culture of centralized, top-down management, with the other (non-JPL) field centers operating under tight authority. This was not, at the time, how JPL rolled.

JPL engineers grumbled about the accountants and pencil pushers at NASA, while NASA chafed at JPL's air of arrogant independence, augmented no doubt by their unique relationship with Caltech. It was not a promising start to the project.

By 1962, despite the delays with the early Venus-bound Mariners, JPL was proceeding apace with their design studies for Voyager Mars. NASA headquarters was uncomfortable with the doings on the West Coast and demanded increasingly frequent—and detailed—reports on these activities. What they got was deemed incomplete and insufficient—they felt that they were either not getting the whole story or that the JPL folk were not proceeding in a sufficiently disciplined manner from NASA's perspective.[1]

As this dissonance was going on over their heads, the JPL mission planners kept at it. Of particular concern was the atmospheric density of Mars, any measure of which was at best a guess based on telescopic data. The density, or lack of it, in the planet's gaseous envelope would greatly affect how a lander would be designed. To be forewarned was to be forearmed, JPL thought.

By late 1963, work on Voyager Mars was still moving ahead, with JPL determining the instrument load and mass of the orbiters they wanted to build. Headquarters, on the other hand, thought there was too much being done in-house and wanted more involvement from other field centers and private industry. In addition, they still felt that they were not sufficiently in the loop.

Part of the complication of the project was that it was a bit over the top with regard to goals—the program appeared to be fixated on pursuing the search for life on Mars as early as possible. This was a tall order, given what little was known about prevailing conditions there.

While this was proceeding, NASA was looking at other places to build their Voyagers. Other field centers were considered, and at least seven aerospace contractors were conducting internally funded studies of Voyager Mars.

Don Hearth, in charge of the project at NASA headquarters, demanded monthly reports, quarterly reviews, and even the recorded minutes of spacecraft study group meetings from his counterparts at JPL.[2] The Pasadena facility chafed at the extra paperwork—this was not how they did things. Paperwork was necessary to properly manage a project, of course, but the bulk and frequency of all these reports was going to slow things down, and in JPL's eyes it was unnecessary. Many thought that this project would never actually be built but was instead being buried in paper. Hearth, in the meantime, was looking to spend an appreciable amount of money to fund the contractor studies of the project, as was being done all over the country for NASA's Gemini and Apollo programs. This was business as usual for headquarters, as NASA was not considering just the manned lunar landings but a bevy of follow-on human spaceflight projects.

JPL, for its part, felt that they could spend the Voyager Mars dollars most responsibly right under their own roof.

Nevertheless, looking ahead to launch opportunities coming up in 1967 and 1969 (the planetary alignments for Mars come in two-year cycles), NASA selected two contractors to move ahead with more detailed studies of Voyager Mars (NASA just couldn't have too many studies in those days—and, some might argue, now). The AVCO Corporation and General Electric's Missile and Space Division would receive the Voyager study contracts. Despite this, throughout 1963, JPL soldiered on, with an emphasis on a Mars orbiting research platform, with a lander to follow once more was learned about the planet.

In that same year, scientists had ascertained from telescopic data that the atmospheric pressure on Mars was far less than had been thought—possibly by as much as two-thirds. This threw another monkey wrench into the planning—landing would be far more difficult than originally envisioned.

As the year wore on, the spacecraft designs done by the aerospace contractors were getting heavier, while JPL was actually struggling to reach the mass specified for the probe to fly on the desired rocket, a Saturn IB, with a Titan as a backup. They wanted to do orbiters first, which would weigh just five hundred pounds, and felt that the Saturn rocket was overkill for this project. The contractors, on the other hand, had no problem adding weight and complication to their versions of the spacecraft—which could also pad their profits. And they would include landers on the first flights.

AVCO and GE completed their studies, submitting to NASA logical designs for huge Mars landers with a full array of instrumentation, including rudimentary biological investigations. AVCO suggested one lander, and GE two. They were quite different in design—AVCO's was what somewhat like Soviet unmanned probe designs in its approach, using an aeroshell for atmospheric entry, a parachute to slow the craft, and a shock-absorbing, rounded, crushable aluminum pad to blunt the impact. It would roll to a stop and flower-like petals—similar to the ones the Russians had used on their lunar probes—would unfold and force it upright. GE's design consisted of a blunt sphere cone, which would enter pointy end down, like a nuclear warhead, and be slowed by rockets. A system of explosive anchors and levering bars would right the lander once on the ground—ambitious stuff so early in planetary exploration, and it would likely have been ineffective had it been tried.

The designs were intelligent, but, as it turned out, unworkable. In 1965 the news came in from Mariner 4 that the atmosphere was in fact just one-hundredth that of Earth's, and *everyone's* descent and landing designs would have to be reevaluated. It is interesting to note that during most of 1964 JPL had essentially sat on its plans for the Voyager Mars. The history gets a bit murky here—there may have been a clear suggestion to JPL to discontinue work on Voyager, but it seems more likely that the project simply languished at the Lab. Challenges included short staffing, mixed signals from upper management at NASA, and—this is anecdotal—they may have been awaiting the results from their soon-to-fly Mariner 3 and 4 Mars probes regarding the atmospheric density on Mars. If it was as low, or lower, than their own researchers were estimating, any lander designs would not be worth their weight in sand, red or otherwise. As it turned out, they were right.[3]

JPL continued to urge caution, stating that the Deep Space Tracking network, their series of three large radio dish antennas spread across the globe and used for deep-space missions, was not ready to control a complex Mars landing mission. They also felt that research on proper sterilization of landers on Mars would take time, and that there was still work to be done on the nuclear power supplies that were being specified in some of the designs. And in hindsight they were correct—none of these areas had been fully wrung out, and it's doubtful that anyone would have made the 1967 deadline, especially given the all-out dash the Apollo program was making to land men on the moon. NASA already had its hands full . . . but you would not have guessed it by the demands made by Hearth at headquarters.

Voyager Mars's budget had climbed to a then unthinkable $1 billion, and would later be estimated at over $2 billion. Hell, Apollo was only going to cost about $20 billion (so they thought at the time; it was a bit more) . . . but *two billion dollars* to land a robot on Mars? Frankly, it is likely that it would have been more than that.

In the meantime, JPL pressed on with the Mariner flights—Mariner 5 was prepared for a Venus flyby, and Mariners 6 and 7 would fly past Mars. As it turned out, the next Venus flight would occur in 1967, and the subsequent Mars flybys would occur in 1969, two full years after Hearth's desired date for the far more complex Voyager Mars.

After the bad news about atmospheric density from Mariner 4, the Voyager Mars project began to morph into something even more unpleasant. Like a retired quarterback gone to seed, it started to gain more weight at an alarming pace—exactly the opposite of what should have been occurring. Part of this was understandable—the engineers realized that it would need more robust landing propulsion, and a larger parachute, and stronger landing gear, to survive the descent into the ultra-thin atmosphere. By the same token, the spacecraft still needed a weighty heat shield because at the speeds of atmospheric entry the craft would still need to withstand quite a bit of heat. Voyager Mars just couldn't catch a break.

At this point, the planned Voyager Mars system was now not only too large to launch on the rockets that JPL traditionally used—the Atlas-Centaur or Atlas-Agena—but it was even too heavy to be lofted by the Saturn IB, the choice of NASA headquarters and the contractors. So, in late 1965, JPL received news that the new plan was to launch two Voyager Mars probes aboard a single Saturn V rocket. The orbiter/lander combinations had grown from JPL's original estimates of around a ton to *twelve tons each*. The bloated machines were behemoths and, in the opinions of most of the JPL folk, the program was heading off into la-la land.

Fig. 13.3. The monster that was a late-stage design for Voyager Mars. Here the spacecraft, with a lander inside the conical capsule, detaches from a Saturn V upper stage. Image from NASA.

This not only added a huge amount of cost to the project but meant that a single launch failure would doom both spacecraft—redundancy was one of the reasons JPL tended to launch missions in pairs in the first place. But when pitching the "improved" Voyager Mars program to congress in 1967, NASA's associate administrator for space science and applications, Homer Newell, wrote effusively:

> Successes already achieved in the 1960s with unmanned spacecraft of limited weight and power . . . foretell the great work of exploration that lies ahead. . . . With Voyager, the US capability for planetary exploration will grow by several orders of magnitude. . . . Voyager could well be the means by which man first learns of extraterrestrial life.[4]

It was pure hyperbole, and destined to fail. Mariner 4, which had accomplished much at Mars in its brief encounter, weighed 574 pounds. Mariner 9, which would later orbit the planet in 1971—and therefore needed breaking rockets—tipped the scales at 2,200 pounds fully fueled. Even the massive Viking orbiter/lander combination, one of the larger planetary probes of the 1970s, only weighed about 6,400 pounds . . . and the really heavy Soviet Venera spacecraft, built to withstand crushing pressure—about one hundred times that of Earth—and heat up to nine hundred degrees Fahrenheit, were less than 10,000. The Voyager Mars monstrosities were nearly 12,000 pounds *apiece.* And at nearly twenty feet in diameter, they neared the maximum that the Saturn rocket could accommodate.

The JPL people must have wondered who was drinking the rocket fuel at headquarters. Sometimes, under their breath, they still do . . . but that's another story.

And what would you get for all this mass? How about an enclosed capsule design for the lander—strongly resembling the Apollo capsule—with mini-rovers to go forth, procure samples from near the landing zone, and bring them back to the lander for geophysical and life science analyses. Who knew what might be discovered?

For what it's worth, the total mass of the science package would have been about 300 pounds . . . not too much less than an entire unfueled Mariner 2. How the contractors could arrive at a 300-pound science package with a 12,000-pound spacecraft design is something best left to analysis of the blueprints. But that science package was one-fortieth the total mass of the spacecraft—not a very efficient design when compared to something like the Viking landers, in which the instrumentation represented about one-fifth of the total mass. There's a bit of sloppiness in these figures, given the variables of the weight of fuel after touchdown and whether or not you consider a robotic arm to be an instrument, but you get the idea. Voyager Mars was a monster.

Fig. 13.4. By late 1966, Voyager Mars needed a serious dose of Weight Watchers. Instead, Congress killed the program. The much slimmer, and more intelligently designed, Viking program was the successor. Image from NASA.

All these "advancements" would also mean that the launch would be delayed until 1973. A delay of six years does not seem like much in today's cash-strapped planetary program, but during the space race, when we were trying to match and, hopefully, outstrip Soviet accomplishments every few months, an extra six-year wait was unthinkable.

After a tortured and tumultuous NASA-splitting seven-year history, the Voyager Mars program was put out of its misery at the end of 1967. Congress, which was already slashing NASA's manned spaceflight budget (and would later sit by while Apollos 18, 19, and

20 were canceled, along with all the Apollo follow-on mission plans), had endured enough. The free-spending era of American space exploration was over.

There was another culprit beyond the overstuffed nature of Voyager Mars that led to its demise: bad timing. In addition to its other requests, NASA's Houston field center, the Manned Spaceflight Center, put out a request for proposals from the aerospace industry for plans to send astronauts to Mars or Venus for a sample-return mission between 1975 and 1982. Industry was all too delighted to respond to this potentially massive contract . . . but then a member of Congress spotted it. Joseph Karth, a Minnesota Democrat who had been supportive of the Voyager Mars program up until that moment, said that he was "absolutely astounded" by the request, given that NASA had been warned to stay on track with the lunar landings and not start any new programs. And he blamed NASA across the board—the Voyager Mars budget being requested was just as offensive to him as the manned Mars mission request. Karth said, "Very bluntly, a manned mission to Mars or Venus by 1975 or 1977 is now and always has been out of the question—and anyone who persists in this kind of misallocation of resources at this time is going to be stopped."[5] They were, and Voyager Mars died along with any real hope for human missions to the red planet in the Apollo era.

And that could have been the end of a Mars landing for decades. There were still the Mariners set to fly past the planet in 1969, but as of 1968 they were the last funded robots for solar system exploration. Mariner 9 was planned to go into orbit around Mars in 1971—at a budget-friendly price of about $137 million—but had not yet been funded. But, to the surprise of many, NASA managed to pull a rabbit out of the congressional hat, and for the 1969 budget received approval for the slimmer Viking Mars landing program. This effort would see two smaller, more streamlined orbiter/lander combinations launched on two cheaper Titan rockets for a landing in 1976.[6]

It should be said that a number of Viking managers later said that the advance work done for the Voyager Mars program helped them to design the later Viking mission to Mars, so perhaps not all the money spent over the seven-year life of Voyager Mars was wasted after all.

Oh, and the cost of the Viking program? Also about a billion dollars. But at least it worked, and brilliantly.

CHAPTER 14

. . . AND THEN CAME VIKING

Nineteen seventy-six was a good year. The world's first supersonic airliner, the Anglo-French Concorde, was finally flying; Nadia Comaneci would earn the first perfect score in Olympic gymnastics; and punk was just emerging onto the music scene (whether or not that was *good* news depended on whom you asked, of course). In space news, NASA had officially unveiled its plans for the new space shuttle. But for many, topping all of this, was hot anticipation of the first robotic landing on Mars by NASA on the evening of July 20.

I was in my second year of college, studying astronomy without any certain goal in mind. But the subject was fascinating, and as a lab tech at the local college's astronomy department I had every opportunity to spend the warm summer nights with a telescope staring at Mars. My interest in all things space had been sparked by the Gemini and Mariner flights, and I was hooked, so of course I followed the development of the Viking Mars lander program closely. After all, JPL was just up the street from my home . . . and maybe, if I worked hard enough, I'd snag a job there in a few years. It ended up taking many decades, but not through any fault of my astronomy professors.

That warm July night was the seventh anniversary of the landing of Apollo 11, a pleasant conjunction with the landing of Viking 1. This concluded weeks of the lander waiting in orbit—it had arrive a month earlier—while the science teams at JPL fretted over the landscape they saw below. It was far from a welcoming view.

The twin Viking spacecraft, the intelligent successors to the confusion of Voyager Mars, were well-designed and highly capable. There had been struggles in the program, of course—it was a NASA undertaking, after all. At one point, money was so tight that NASA headquarters suggested the deletion of the expensive high-resolution color cameras designed for the orbiters and urged JPL to use those that had been designed for the Mariner 9 orbiter in 1971—medium-resolution black-and-white cameras. Or they could simply choose to not take cameras along if that's what they preferred. Not surprisingly, the JPL people didn't like that. After all the yelling was over, Viking got its improved color cameras, but it had been a fight.[1]

The Apollo lunar landings had concluded in 1972. The new space shuttle was now on the drawing boards and being planned for a maiden launch in the late 1970s at what was projected to be a bargain price (in the end, it wasn't). The heavy expenditures of the Apollo program had begun to drop in the mid-1960s, and now that the manned lunar landing program had drawn to a close, there was more money available for planetary exploration. But you would not have guessed it when talking to the people working on Viking—they still had to fight for every penny. But that's how it goes in space exploration, now as then.

While ultimately costing as much as the early and optimistic projections of the Voyager Mars program, the Vikings were the near-perfect Goldilocks combination of cost and capability. These were twin orbiter/lander combinations. The Viking spacecraft were still large, weighing in at a combined 6,400 pounds, but they were an ingenious design. The orbiters were upgraded Mariners at their cores, and the landers drew from Surveyor's DNA. Both were built through the combined efforts of NASA's Langley Research Center, JPL, and the Martin Marietta corporation, which was later absorbed into aerospace giant Lockheed.

Each lander had a pair of high-resolution (for the time) color cameras aboard that could take 3-D images of the Martian surface. Each also had a robotic arm for procuring samples of nearby soil, which would then be delivered to an onboard, automated laboratory for analysis. And while understanding the minerology of the planet was highly desired, the onboard labs were after something even more delicious: life.

Fig. 14.1. A technician checks out (or more likely, poses with) the Viking lander's soil scoop on the end of its ingenious robotic arm. Image from NASA/JPL-Caltech.

Tucked inside each ten-foot-wide by seven-foot-tall lander was a miniaturized life-science laboratory. This set of instruments was a marvel for the 1970s and would search for microbes in Martian soil (more properly called regolith), which would be delivered to them by the arm-mounted scoop. But how to conduct the analysis of that soil was a problem that consumed the small group of scientists assembled to design an instrument package that would weigh less than forty pounds, much less than any equivalent individual instrument would weigh on Earth.

This package included four experiments. These were not optical microscopes or other direct-observation instruments but, rather, cleverly designed devices that would seek out indirect indications of microbial life—the only kind of organisms that could be envisioned by the biologists in the planetary science community after the results of the Mariner observations of Mars. There was still not a lot known about the surface of the planet, but certainly much more than was known before Mariner 4 flew by. Prior to that, when the first designs for the life-science lab on Voyager Mars was being designed, the scientists could have been looking for anything from lichen-like plants to Mars cows (however unlikely the latter might have been). Now that the scientific community had at least a basic working knowledge of Mars—dry, cold, and with a very thin atmosphere composed primarily of carbon dioxide—they had at least a starting point from which to proceed.

There were a number of scientists involved in the life-science experiments aboard the Viking landers. Each experiment was comprised of a small, airtight container in which a sample of Mars soil would be deposited and subjected to some kind of intervention, and the results observed and interpreted by the science team.

One instrument was called the Pyrolytic Release (PR) experiment. Inside this chamber, soil would be exposed to light, injected with purified water, and exposed to carbon dioxide (CO_2) and carbon monoxide (CO) infused with slightly radioactive carbon 14 (^{14}C). This would, in effect, expose the soil to a controlled version of what the Martian environment was thought to be like. After a few days, the chamber would be evacuated, then heated to high temperatures. The resulting gases that were baked out of the sample would be examined to see if there was a radioactive signature attached to the remaining gas, which could indicate that it had been ingested and metabolized by Martian microbes.

The second instrument was called the Gas Exchange (GE) experiment. This device would first purge the Martian air in the chamber by injecting helium, which is inert, and would then add nutrients into the soil, thought to be universally attractive to bio-organisms for ingestion. This process would be done first with just the nutrient broth, then with both the broth and purified water added. A gas chromatograph would then be used to analyze any output from the soil, looking for signs of metabolism by microbes—oxygen, carbon dioxide, nitrogen, hydrogen, or methane.

The third instrument was called the Labeled Release (LR) experiment. Like the PR experiment, Martian soil in the chamber would be injected with nutrients that contained radioactive carbon 14 and observed for signs of metabolism by searching for carbon dioxide with a ^{14}C indication, demonstrating possible metabolism by organisms within the soil.

The last instrument was a combined gas chromatograph and mass spectrometer. When this soil sample was enclosed in its container, the gas chromatograph would separate any vapors from the soil, then feed them to the mass spectrometer for analysis. The mass spectrometer would be able to identify the molecular weight of each constituent, looking specifically for the biosignatures of organic molecules that could be indicative of life. This was the only instrument that did not require the "incubating" of any life forms that might be squiggling around in the soil sample—it merely examined what was there when the sample was collected.

It is noteworthy that all the experiments except for the last depended on Mars organisms to behave in ways that paralleled Earthbound equivalents as they were known in the late 1960s when the Viking experiments were designed; but this is how planetary exploration works—one builds on assumptions of science as it is understood when an exploratory program is being created. New probes, such as the Mars 2020 rover being prepared to launch later in this decade, proceed with different assumptions, based on what is known now. But in 1976, this was state of the art.

Both of the Viking landers carried identical sets of instrumentation, and were identical in all respects.

Of the scientists involved with the life-science experiments, the two names that got the most public attention were Norman Horowitz, who designed the PR experiment, and Gilbert Levin, who was responsible for the LR experiment. The two scientists had a difficult relationship, based on some past history and the results of their experiments on Mars. In sum, Horowitz was at best a cautious optimist, almost a doubter, and Levin was an optimist with regard to possible Martian life. The results of their experiments exacerbated these tendencies, resulting in a number of cantankerous exchanges once the preliminary results were in, and long afterward.

Horowitz was in the driver's seat, however, since he was the chief of JPL's biosciences division. His ideas were clear from the beginning: "I thought [life on another planet] was a plausible idea. Everything that was known about Mars at that time later turned out to be wrong, but [at the time] suggested that there was a good possibility of life on Mars," he said. "JPL [had] set up a bio-sciences section to plan for the biological exploration of Mars, with an eventual lander. They asked me to come up and be chief of their section, which I did in 1965. There was a lot of work going on up there in trying to design instruments to fly to Mars for a biological search, and I got involved in that planning."[2]

One of his primary goals was to assure that there was ground-based experience with the instrumentation before committing it to flight—he wanted to make sure that any experiments carried on board were based on known, terrestrial science. Speaking of his own experiment, he said,

> The instrument is based on empirical patterns of breakdown of organic compounds. You take an organic compound and you heat it until it pyrolyzes—it breaks into smaller fragments due to the heating. These fragments can be identified by a combination of analytical steps called gas chromatography and then mass spectrometry. The only thing you have to identify the original compound you started with is the pattern of its breakdown products, and you try

to infer the nature of the original compound from these breakdown products. There's not much general principle of general theory you can go on; you just have to have a library of results you can compare your actual results with.[3]

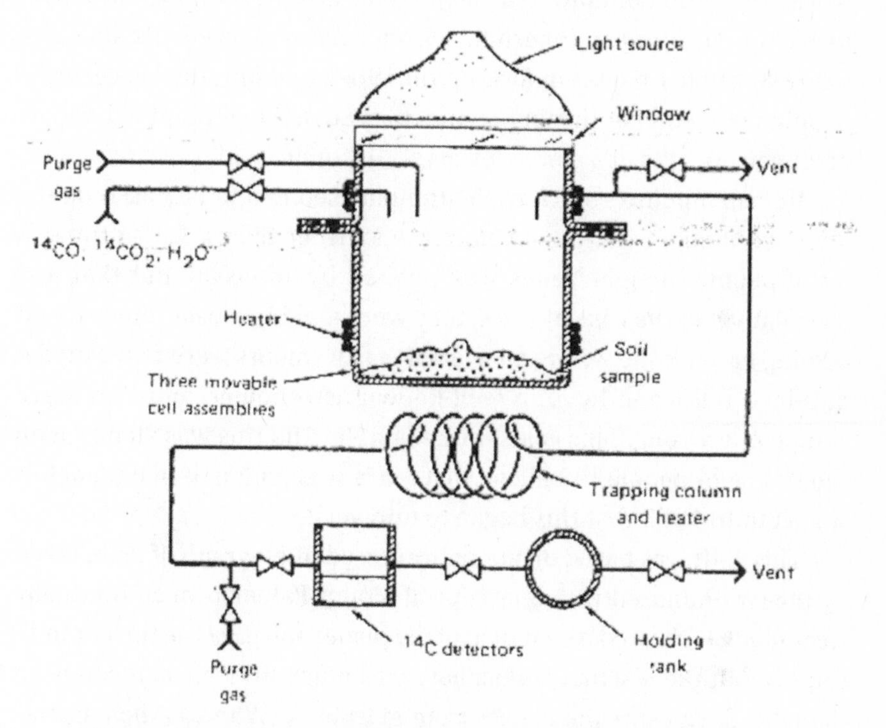

The pyrolytic release experiment put radioactively tagged ^{14}CO and $^{14}CO_2$ gases with a sample in artificial sunlight, let it incubate for several days, and then heated it to see if tagged carbon had been incorporated by the sample. It was, and markedly more than in a "sterilized" control sample.

Fig. 14.2. Norman Horowitz's PR experiment would soon come to represent a pitched battle between Horowitz and Gilbert Levin, who was responsible for the LR experiment. They both saw the same readings, but their interpretations differed—a lot. Image from NASA.

By a "library," he meant the assembled results of similar studies conducted on Earth.

Horowitz was wary of some of the motivations behind the overall life-science package traveling to Mars. "It's hard to convey in a few words the total commitment people had in those days to an Earth-like Mars. This was an inheritance from Percival Lowell. It's amazing: in pre-Sputnik 1 days—in fact, up till 1963, well into the space age—people were still confirming results that Lowell had obtained, totally erroneous results. It's simply bizarre!" Horowitz said.[4]

He had a point—much of the thinking seemed to be based on the Mars we thought we knew before the Mariner flybys. At the time, "A lot of people thought Venus was covered by an ocean. But that was speculative; in the case of Mars, they were making measurements and coming up with the wrong answers. Measurements were made on the 200-inch telescope by . . . a well-known astronomer, and they were completely wrong. This is just one example. And this was all based on the desire of people to believe that Mars was an Earth-like planet. It wasn't until 1963 that this began to unravel."[5]

This shift was based on another infrared photograph of Mars taken by the two-hundred-inch telescope at Mount Palomar on an unusually clear night. It showed a heat map of the planet and gave the further indications that the Martian atmosphere was much thinner than had been thought. Then came the results from Mariner 4. Whereas the Palomar image implied an atmospheric density of perhaps 85 millibars—less than a tenth of Earth's at sea level, which is just over 1000 millibars. "Mariner 4 flew by Mars and found that the surface pressure was more like 6 millibars . . . and that is the average pressure. And carbon dioxide is the principle gas in the atmosphere. Well, with 6 millibars, there's virtually no chance of having any liquid water," Horowitz said.[6]

This was crushing news, since it meant that there was a lot more carbon dioxide in the Martian atmosphere than had been assumed, and that there was little chance of water existing in the liquid form scientists thought was needed for metabolism by life. In subsequent decades, other forms of life have been found on Earth, called extremophiles, that don't require water found in liquid form to survive; but in the 1960s, this was not understood.

Horowitz continued:

> In spite of all these new discoveries, people were still building instruments to fly to Mars that were based on the terrestrial environment, and they were eventually approved by NASA. NASA was supporting these efforts. Around 1960, I got involved in one of them, one that actually later flew on Viking. We called it Gulliver at the time. It was invented by an engineer in Washington named Gilbert Levin. It depended on an aqueous medium [aka liquid water]. Two other experiments that were being supported by NASA also involved aqueous solutions into which you would put the Martian soil and then use various ways of measuring the metabolism of the organisms. But after 1965, after the Mariner 4 flyby, it was obvious that the chance of liquid water on Mars was so remote that one had to plan for the contingency that there was no water—that if there was any life on Mars, it was living under conditions that were in no way terrestrial. So we designed an experiment that would work under Martian conditions and that involved no liquid water.[7]

Horowitz and Levin were on a collision course over the question of life on Mars. But first, Viking had to get there. Viking 1 roared off a Cape Canaveral launch pad in August of 1975, and Viking 2 followed in September. Viking 1 arrived at Mars orbit on June 19, 1976, and Viking 2 on August 7, 1976. Unlike modern Mars probes, they braked into orbit. This was for two reasons: first, this allowed the scientists on Earth to scope out possible landing zones for each lander before committing to a landing spot below. Second, the orbiters would remain in orbit around Mars to image and map the planet for years to come.

There had been much acrimony over where to place the two landers. The orbiters were easy—just sling them into orbits from which they could image the planet below and release the landers when the time was right. Viking 1 was put into an orbit inclined about forty degrees to the Martian equator, and Viking 2 at about eighty degrees. But where to send those landers was a tougher question, especially considering that finding life was a top priority of the mission. The poles were an obvious consideration—there is water ice there, though these regions are bitterly cold. Mariner 9 had identified what appeared to be water-carved features closer to the equator—would those be a good place to look? Since it was assumed that any water that had flowed at these sites did so in the distant past, perhaps not. Other regions with different qualities were examined and discussed, with pros and cons hotly contested.

Leading up to the mission, the team responsible for preliminary Viking landing site decisions organized what they called the "Planetary Patrol." Working with facilities like Lowell Observatory in Flagstaff, endless observations of Mars were made and reams of data examined in an effort to track cloud formation, dust storms, and anything else that could be gleaned from the limited abilities of Earth-based observing. They even engaged teams of graduate students—the cannon-fodder of much scientific inquiry—to do crater counting. The kids would pore over thousands of images from Mars orbiters and flybys, counting craters in a specific grid pattern, in an effort to infer how rough the surface might be. But this was a bit like trying to find an individual pinecone by looking at a forest from sixty thousand feet in the air with the naked eye and guessing which trees might have pinecones, with only large clumps of trees being visible. On Mars, any rock much larger than an over-inflated soccer ball, or crater rim over a foot tall, could ruin a Viking lander, or at least keep it from operating properly—the machines had only about ten inches of ground clearance. In the end, such exercises were ultimately marginal in effectiveness, but the exercise made everyone feel better—they were doing what they could.

Prior to the arrival of the Vikings, the most recent batch of Mars images had come from the Mariner 9 orbiter in 1973. While these were a huge improvement over the previous Mariners—all of which had been flyby missions—the smallest details that could be discerned were still over three hundred feet across.[8] This had given mission planners a general idea of the quality of the terrain, but it was insufficient for choosing landing sites for the two Vikings—if they came down in a field of boulders, or on the rim of a crater, the lander's mission could be over before it started.

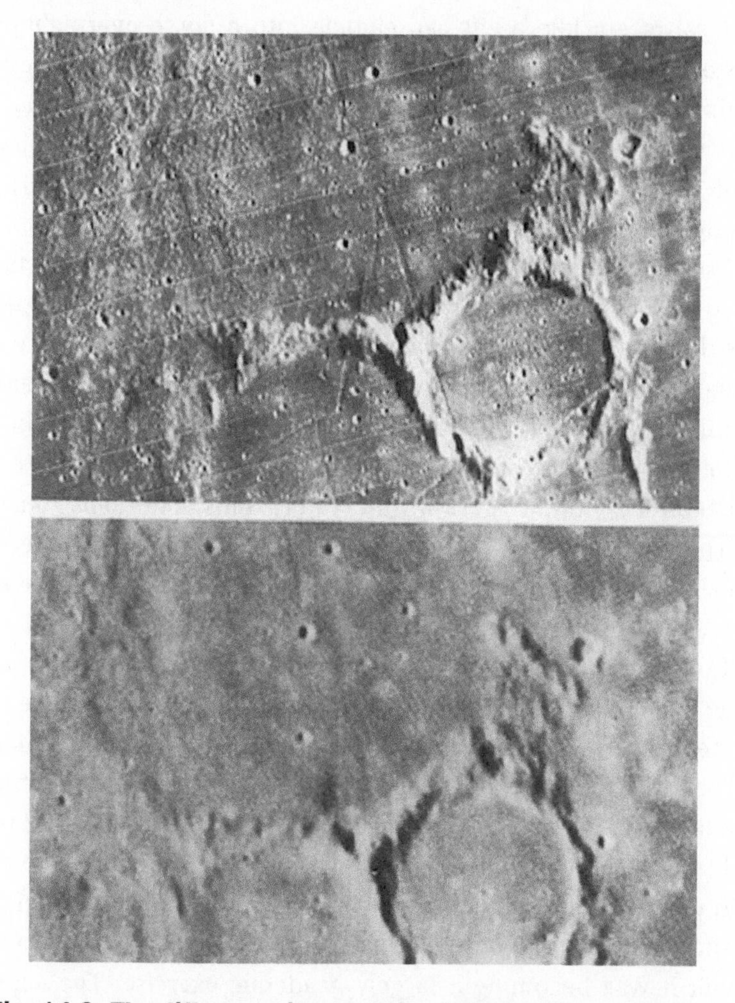

Fig. 14.3. The difference in resolution between the cameras aboard Mariner 9 (bottom) and the Viking orbiter (top) are obvious. It's no wonder that a) the JPL engineers threw a fit when NASA tried to "descope" (or shrink the budget and complexity of) the Viking cameras and b) the same engineers shuddered in their shoes when they saw the new images of Viking 1's landing zone . . . the enhanced detail was frightening. Images from NASA/JPL-Caltech.

The Viking orbiters, on the other hand, were able to image areas as small as twenty-five feet across—far better than the Mariners. But what can seem like a gift can change into a curse overnight; what mission planners saw shocked them. The surfaces of the proposed landing zones were far rougher than anticipated, and even when they were able to identify relatively smooth, unblemished areas, at a resolution of twenty-five feet, there could be all kinds of nasty surprises lurking below.

When looking at the earlier photographs, many geologists surmised that, while Mars had some larger craters, it did not appear to have the extremely rough surface seen on the moon. They were wrong. As the images of Mars improved in resolution, what greeted them was not unlike what they had seen in the Ranger photos of the moon—the closer you got to Mars, the more craters you saw. There were once again craters within craters within craters. This made sense once you incorporated how thin the atmosphere was now known to be, since thin air cannot incinerate meteorites as effectively as a thicker atmosphere. But it was still a shock.

Both imagery and radar readings from the Vikings were added to the mix, until at last, with much trepidation, the landing site selection teams came up with two areas in which to drop their expensive spacecraft: Viking 1 would land in a region called Chryse Planitia (the Golden Plain), which was near the equator of Mars, and Viking 2 would land farther to the chilly north, in a region called Utopia Planitia (the Plain of Paradise). The debates on exactly where in these regions to set the machines down raged on until the final days before landing, though it was becoming a largely academic exercise. The expected landing zone could only be loosely defined anyway—the machines would be descending autonomously, since Earth was almost twenty minutes away by one-way radio, and they would land where they landed—the aiming by mission control was far from precise. The landing zone, called a landing ellipse, measured 174 by 62 *miles*. This was the projected area within which the landers would touch down . . . somewhere. So any smaller areas they preferred for a touchdown were at best an approximation. But that did not stop the scientists from arguing.

On the other side of the discussion were the engineers. While the life-science people wanted areas that were likely to provide the best chances of biological activity, and the geologists wanted places where there were likely to be a wide variety of interesting rock and soil specimens to investigate (like a dried-out river delta, for example), the engineers just wanted something *safe*—a nice, flat plain would do nicely; something akin to a dry lake bed, like those found in the high deserts of California and Utah, would suit them just fine. Flat and smooth was what they wanted—as few boulders and craters as possible, please. But for the scientists, these regions were likely to be just hard, basaltic plans, the result of lava flows later in Mars' geological development, with hard surfaces into which the robotic scoop would be unable to dig, and so they were not a prime choice. The arguments raged on. The landing, which was originally scheduled for July 4 and was intended to coincide with the Bicentennial celebration then ongoing in the United States, slipped almost three weeks as the landing sites were analyzed and agonized over. And then, it was time to go.

On July 20, 1976, Viking 1 detached from its orbiter and began the long, three-hour sequence of events that would hopefully put it down on the Martian surface in one piece. Remember that this was all autonomous, using 1970s computer technology; these machines had eighteen kilobytes of memory and could perform about 25,000 instructions per second. This is an infinitesimal amount compared to what we use daily today—your old flip-phone was a processing beast in comparison—but it was what they had to work with. With this limited computing power, the lander had to assimilate data from onboard acceleration sensors and a downward-looking radar unit, then combine this with its onboard data of speed and orientation to make a landing decision. It was a tall order, and a daring undertaking for 1976, a year when high tech in Detroit still meant the Chevy Chevette and the Ford Pinto, and Ma Bell (the national telephone company) was just coming out with push-button phones!

Fig. 14.4. Artist's impression of the Viking orbiter/lander as it nears Mars. Mariner's DNA can be clearly seen in the orbiter section, the lower part of the spacecraft with the flag. More complicated and expensive was the lander, in the conical aero-shell at the top. Image from NASA.

Of course, I knew little of the drama playing out at JPL in 1976. I just understood that two amazing machines were headed to the surface of Mars, and I wanted to be a part of it in any way possible. Access to JPL was not available to run-of-the-mill astronomy students, and so that particular temple of planetary exploration was not open to me. But I knew that the mission would be closely watched at Caltech, just down the street, and I endeavored to find a place there from which I could observe the proceedings.

After a number of phone calls, it was clear that this was not an option either—access was limited. Fortunately, I'd grown up in the area, and many youthful hours were spent exploring the Caltech campus, including, shall we say, less-known and little-patrolled entrances to the auditoriums on site. I arrived there in the late evening for an expected landing of just before 5:00 a.m. Pacific time. After entering one of Caltech's auditoriums via somewhat unauthorized means, I settled into the back of the room with little notice, apart from a few jaded reporters who surely noted my lack of press credentials. I could have been a Caltech student, so long as they were not privy to my SAT scores, but a couple of them still smirked knowingly. I shifted about uncomfortably in the darkened room.

On a large screen at the head of the auditorium were images of JPL's landing team monitoring the events already unfolding many tens of millions of miles away. Everything we saw was delayed by about eighteen minutes, so it was a bit like watching the World Series after the fact—anything you witnessed had already occurred, and the news was entirely past tense. Prior to this, JPL's only experience with delays in off-Earth landings was with the Surveyor, with a delay of about 1.2 seconds between the Earth and moon—hardly enough to notice when compared to eighteen *minutes*.

In the wee hours of the morning of July 20, after completing its disengagement from the orbiter, the Viking 1 lander fired rockets to slow down and descend toward the surface of the planet below. It oriented itself heat shield first—the first time NASA had tried this beyond Earth orbit—and made its fiery entry into the atmosphere. Despite being just a fraction of Earth's atmospheric density, the Martian atmosphere is still thick enough to cause a great deal of heating, just not sufficient for parachutes to fully slow a spacecraft—it is, in a sense, the worst of both worlds.

At about four miles' altitude, an enormous single parachute deployed, slowing the lander's descent. Then, at about five thousand feet, the landing rockets were fired by a signal from the descent radar. These rocket engines were of an ingenious, throttleable design, with an array of eighteen tiny nozzles to minimize the effects of the exhaust plumes on the soil at the landing site—they wanted anything sampled to be as pristine as possible. Viking 1 set down at the pace of a brisk evening stroll, about four miles per hour.

Back at Caltech, we were still watching the final phases of the descent—nobody knew if the lander had set down gently or crashed in the attempt 225 million miles away. Then, after an agonizing wait, with only announcements of what *should* have occurred streaming in, we heard the excited voice from mission control at JPL: "Touchdown! We have touchdown . . . we're looking good." Cheers erupted at both the Lab and at Caltech—many people on campus had invested the better part of a decade into the project, and the tears flowed freely.

Gentry Lee, the solar system chief engineer at JPL, later said, "The Viking team didn't know the Martian atmosphere very well, we had almost no idea about the terrain or the rocks, and yet we had the temerity to try to soft land on the surface. . . . We were both terrified and exhilarated. All of us exploded with joy and pride when we saw that we had indeed landed safely."[9]

Then, as the narration from JPL continued, the auditorium grew quiet again. Everyone was waiting for that first image from the surface. Soon, it began to come in . . .

This first photograph was slow to arrive, forming one vertical strip at a time. The cameras on the Viking landers, mounted to the top deck of the spacecraft, looked like two coffee cans with vertical slits cut into the sides. Inside was a mirror mounted on a pivot, which would scan up and down, feeding the image to a camera below. Strip by strip they arrived, moving left to right. Video projection was not then what it is today, and the projected image at the front of the auditorium was purple-hued ... but nobody cared. This was the first image taken from another planet's surface to come back to Earth.

But within minutes there was a low grumble from the press. "Where's the horizon?" asked one crusty reporter to my left, holding a notepad with a pencil suspended above it in anticipation of the moment. For what had just been unveiled to us was an image, not of the Martian horizon but of a patch of Martian dirt. One of Viking's footpads was revealed on the right side. More grumbling. "They're looking at its *feet*?" the wag continued, incredulous.

Indeed, they were.

"We wanted to understand what the bearing strength of the surface was. We wanted to understand how the surface would respond to the footpad," said John Newcomb, a member of Viking's mission management team, years later.[10] It was important for the engineers to make sure that the lander had arrived safely, was firmly planted on the surface, and would not be sliding down a hill or sinking out of sight in some equivalent of Martian quicksand.

Press hacks aside, the moment could not have been more glorious. And it served, in a way, to heighten the anticipation of the next black-and-white image to come down from Viking, that of the horizon. Within hours, JPL convened a press conference to address the obvious question: had they seen anything to indicate the presence of life on Mars?

"There are certainly no features in these pictures that have to be due to life," said Carl Sagan, who was assigned to the program as a visiting scientist from Cornell University. "No obvious bushes, trees, or anybody else ..." The clever use of "any*body*" elicited chuckles in the auditorium. "In my view it doesn't exclude the possibility of even larger organisms elsewhere on the planet, any more than landing on a random place on a desert [on Earth] and not seeing anything would give you a reliable extension to the rest of the planet ..."

Gerald Soffen, the Viking project scientist, said, "If you gave such a scene to a biologist in a desert, and told him this was the Earth, and asked him to guess how many forms or organisms might exist there, his numbers would easily go into the hundreds."[11] Hope continued that life would be found on Mars—it could be hiding anywhere, tiny and unseen.

The following day a color image arrived, and humanity finally saw Mars's real estate for what it is: a harsh alien environment with hauntingly familiar elements. It might not look like what we'd hoped for . . . there were no trees, bushes, or *canals*, but seeing from ground level it was still a delight.

In this first color image of the horizon the sky appeared to be an earthly blue and the soil reddish-pink. Once the color had been corrected by JPL's photo lab the next day, it showed a salmon-pink sky and stark orange landscape. There is art as well as science in the discipline of planetary image manipulation, and it took those image-processing experts, pioneers in dealing with color images from space, a while to get it right. They probably expected a pat on the back from the press. If so, they thought wrong.

When this picture was revealed at the next press conference, a few of those in attendance actually booed. One was heard to ask if the sky might be green tomorrow, which elicited an embarrassed smirk from one of the image-processing technicians. JPL had worked the night through to get the colors right, and this was their reward? Nevertheless, it was fantastic, and other members of the press corps shushed the offending member amid some chuckles.

Sagan gave the news: following their further analysis, "The sky is not blue . . ." Boos, catcalls, and chuckles. "It's a typical Earth chauvinist response," Sagan continued, chuckling himself, and was joined by the press. "According to Doctor Pollock," he said, referring to the imaging specialist at the Ames Research Center, "the sky is actually in fact pink . . . which is an okay color."—More laughter.—"The sky is red, but not as red as the surface."[12]

On September 3, 1976, the feat was repeated when Viking 2 set down in Utopia Planitia. The press was less enthusiastic—they'd already seen Mars close up once. But the landscape here was different—far rockier and with more rolling terrain. And . . . it was tilted.

Viking 2 had landed with one of its three footpads planted on a rock. Fortunately, it was a small tilt and did not affect the successful operation of the lander.

As the days went by, the data began to stream in. Martian weather was charted, and the atmosphere analyzed. It was indeed far thinner than our planet's, about 0.09 pounds per square inch, compared to Earth's 14 pounds at sea level. Over time, both landers measured the average temperature to be about -67 degrees Fahrenheit, with a rare high in the 70s and extreme lows in the -200s. At Viking 1's equatorial landing site, short-term measurements recorded a comparatively moderate -22 to -139 degrees Fahrenheit. Wind speeds averaged about ten miles per hour.

Then another glitch, this one a potential mission killer. As controllers slowly extended the sampling arm to begin operations it stuck fast. A latching pin, a safeguard against damage during launch, was stuck in place, and nothing they did seemed to help.

Engineering teams ran over to the Viking clone in the sandbox at JPL. All they knew at that moment was that the arm was stuck; they soon realized that it must be the safing pin, designed to keep the arm immobile during flight and landing. After much testing and practice on the engineering twin of Viking's arm, they managed to shake, rattle, and roll the errant retaining pin out. The process was carefully repeated over a series of days on Mars, and, at last, the pin dropped free. The arm could now extend and retract as designed, and the big moment was upon the biologists: they could now deliver Martian soil to their coveted life-science experiments.

The three experiments with their little ovens, and the gas chromatograph/mass spectrometer, received their samples and went to work. Results were observed and collated. Expectations ran from highly optimistic to cautiously hopeful as the numbers came in. And this is where the discord in the life-science team began.

At first, the results *seemed* to indicate that there might be microbial life in the soil—each of the three life-science experiments exhibited readings that were in line with this possible interpretation. The

Gas Exchange experiment showed a buildup in pressure—something was causing excess gas inside the chamber! Then the Labeled Release experiment demonstrated a small amount of radioactive gas, possibly the result of metabolic activity! Even grumpy Norm Horowitz's Pyrolytic Release experiment was showing some results.

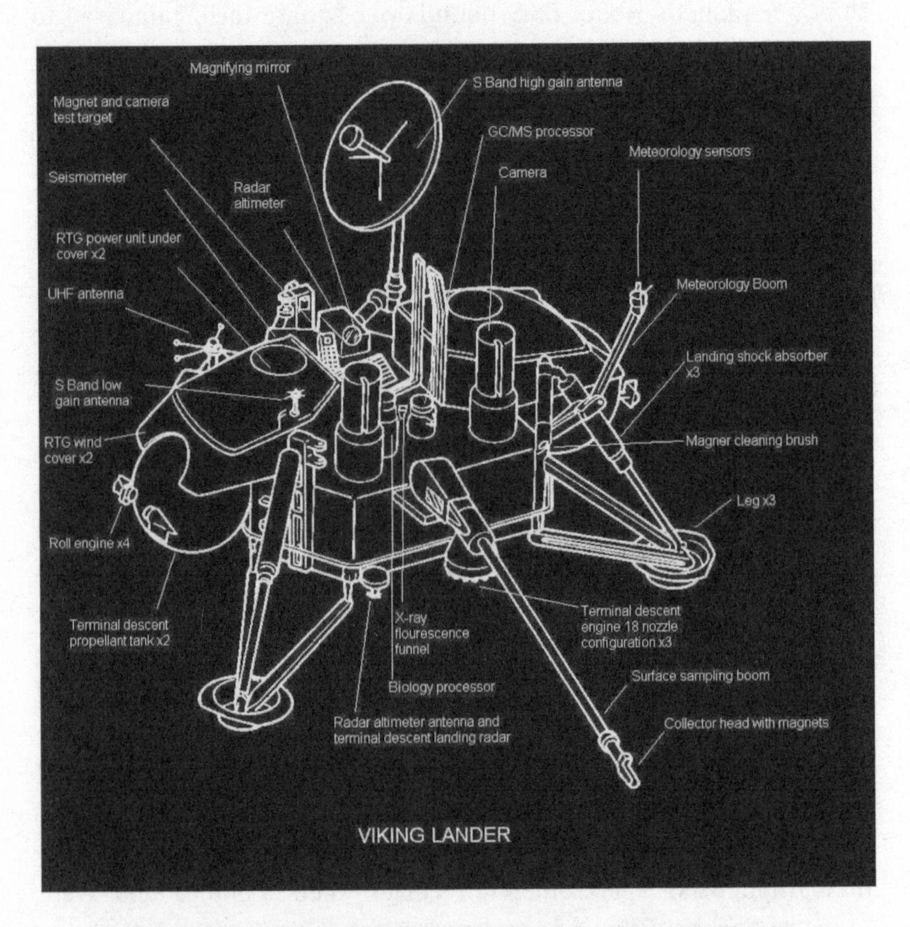

Fig. 14.5. The Viking 1 lander with the arm extended. For the first few days, engineers could only hope that the robotic arm, seen extended to lower right, would eventually work. Image from NASA/JPL-Caltech.

But the pattern of these indications did not match what they would have expected from biological metabolism. The soil would react once, then not react again when exposed to more water or nutrients. And the speed at which the reactions occurred looked more like—reactive soil chemistry?

At a press conference on July 31, just eleven days after landing, the biology team leader, Harold Klein, said this to the assembled press: "What we are proposing to do for you today [is] to give you a status report on the three experiments, and we'd like to then focus on one of the experiments, the Labeled Release experiment, a little more closely since some of that data is exciting and interesting."[13]

First up: the Gas Exchange (GE) experiment had given the team an indication that led them to believe that they might have caught some bugs. "We have at least preliminary evidence for a very active surface material. ... We believe that there's something in the surface, some chemical or physical entity which is affording the surface material a great activity."[14] Here he added a word of caution to the already carefully crafted wording—the GE experiment might be mimicking some aspects of biological activity, but this did not necessarily mean that was biological in nature. Then, he announced that Levin's Labeled Release (LR) experiment's radioactivity counters were measuring "a fairly high level of radioactivity which to a first approximation would look very much like a biological signal." The rate at which the soil was reacting caused some caution, however—it was building pressure more quickly than some thought metabolism would. "That second result must be viewed very, very carefully in order to be certain that we are, in fact, dealing with a biological or non-biological" phenomenon, Klein added.

Just a day later, gas levels were dropping across the board, leading most team members to conclude that they had been seeing a purely chemical, non-biological reaction. But Levin saw something different: while he noted no evidence for the timing of cell division, which might have been expected, the results did not seem to specifically behave as they would have for mere chemical reactions either. "We find that the chemical reaction took place at a very rapid rate initially, and then uncharacteristically slowed down and took a long time to plateau," Levin said at another press event.[15]

Repeated runs at both landing sites were anticlimactic. There were still some puzzling results, but each was far less spectacular than the first outing.

When everything was considered, it appeared that Martian soil was indeed reactive, but not due to biological activity. The hypothesis emerged that the dirt was laced with peroxides that reacted with the water that the various experiments injected into the sample. This would give the types of readings that they were witnessing. The tipping point in this argument occurred when they looked at the results from the gas chromatograph/ mass spectrometer—no organic molecules had been spotted.

But nobody could be certain, and the team divided into the we-may-have-found-life camp and the it-was-soil-chemistry camp. At one point, it got so bad that Horowitz and Levin nearly came to blows at a conference. The debate goes on, and will not be resolved until the mission of the Mars 2020 rover or Europe's ExoMars rover, at the soonest.

There have been indicators, however. In this corner, the pro-Mars-life proponents. And in this corner, the soil-chemistry advocates.

In favor of the latter, the Mars Phoenix mission of 2008 made a confirmed observation of perchlorate (a form of the oxidizing chemicals theorized by the Viking team) in Martian soil. One point for Horowitz.

But wait—experiments in the past decade, attempting to replicate the results of the weird positive indication of Levin's Viking experiment, using soil laced with both perchlorate and bio-organisms, have shown that soil with critters and the oxidizing chemicals could have caused a very similar reading when combined. One point for Levin.

Horowitz died in 2005 at the age of ninety, and he maintained until the end that his conclusions were correct. As he later said,

> There are doubtless some who, unwilling to accept the notion of a lifeless Mars, will maintain that the interpretation I have given is unproved. They are right. It is impossible to prove that any of the reactions detected by the Viking instruments were not biological in origin. It is equally impossible to prove from any result of the Viking

> instruments that the rocks seen at the landing sites are not living organisms that happen to look like rocks. ... The field is open to every fantasy. Centuries of human experience warn us, however, that such an approach is not the way to discover the truth.[16]

Perhaps he'd been watching one of those Star Trek episodes where aliens looked like big rubber boulders. Score one for Dr. Snarky.

Levin is still alive, still working, and he continues to support his point of view—that there was a positive indication of life—a position he published in 1997.

"There was strong opposition to any biological conclusion, based primarily on the failure of the Viking GCMS to detect organic molecules," Levin said. Then came the perchlorate idea, "after which a plethora of variant oxidant theories were put forth until the present." The debate has not gone away, despite the presence of five more landed missions on Mars. "These [theories] were all capped with the insistence by people such as Norman Horowitz ... that there could be no liquid water on the surface of Mars, hence no life," he said. "I followed and refuted all the arguments," he added, " ... until, in 1997, it became obvious to me that, all facts considered, the [Labeled Release experiment] had, indeed, discovered living microorganisms on the surface of Mars."[17]

Notably, just a few years ago, NASA's Mars orbiters spotted dark streaks on some slopes on Mars. These appear to be seasonal, and a theory was quickly put forth that they could be RSLs, science-ese for Recurring Slope Linea—seasonal liquid water that seeps out of rocky slopes. If this is true, it would probably be salty water—a brine—since this evaporates at a lower pressure than fresh water. But a brine is still liquid water, and that would be a huge discovery on this hostile planet.

"Our quest on Mars has been to 'follow the water,' in our search for life in the universe, and now we have convincing science that validates what we've long suspected," said John Grunsfeld, then the associate administrator of NASA's Science Mission Directorate. "This is a significant development, as it appears to confirm that water—albeit briny—is flowing today on the surface of Mars."[18]

Co-researcher Lujendra Ojha, of the Georgia Institute of Technology in Atlanta, added, "We found the hydrated salts only when the seasonal features were widest, which suggests that either the dark streaks themselves or a process that forms them is the source of the hydration. In either case, the detection of hydrated salts on these slopes means that water plays a vital role in the formation of these streaks."[19] In layman's terms, salt deposits detected on the same slopes seem to indicate that they were left there by water that dried out after flowing.

The results are inconclusive, with some researchers subsequently suggesting, despite the indication of dried salts where the dark streaks have occurred, that the dark areas could also be dry landslides of darker soil, but the notion that there may be seasonal water flows on Mars under the right conditions is intriguing.

NASA had planned for the Viking landers to have a primary mission of ninety days, though longer operations would be welcomed. Thanks to their nuclear fuel supply, JPL's brilliant design, and Martin Marietta's robust construction, the four spacecraft lasted for years.

Then, one by one, the Vikings ceased operations. The last to go was Viking 1, which outlasted both of the orbiters and its twin lander. After almost six-and-a-half years of pioneering work, Viking 1's long run ended in November of 1982. It was still sending back weather reports from the Martian frontier and was due for a software update. With its nuclear power supply, it should have been good for at least a few more years. But within the software update, which was designed to help the batteries charge and discharge more reliably, there was an error in the code. The uplinked code apparently overwrote previously existing software that directed the radio dish to point at Earth, and there was no additional "safing" or recovery routine included in Viking's software. The lander did as it was told, and contact with Earth was lost. Despite diligent efforts from JPL, there was no further communication, and that was that. Nobody knows how long Viking 1 continued to stare into the cold Martian sky, awaiting another command that would never come. There may well be some electrical current flowing in its nuclear heart to this very day. We will likely never know.

CHAPTER 15

FLASH FORWARD: MARS HAS MOXIE!

Before we can send humans to Mars for the long term, we will need to work out how to use resources found there to support them. While Mars is the best candidate among our planetary neighbors for potential colonization, it nevertheless has numerous deficiencies, and among them is that the planet lacks breathable oxygen. An experiment that will soon land aboard the Mars 2020 rover should pave the way for producing oxygen from the Martian atmosphere. Charmingly, it's called MOXIE ("moxie" means determination or grit as a character trait, as in, "That Gil Levin sure has moxie!").

MOXIE stands for Mars Oxygen In-Situ Resource Utilization Experiment. Its sole purpose will be to demonstrate the ability to generate pure oxygen electrochemically from the Martian atmosphere. Martian air has abundant carbon dioxide—it's about 96 percent of the planet's atmosphere—and from this oxygen can be created. Oxygen is, of course, valuable for human survival on the red planet, as well as for supplying fuel to returning spacecraft such as a robotic sample-return vehicle. A relatively simple electrochemical reduction is all that is needed to generate oxygen from Martian air, but, while it may be a simple process, it's never been tried on Mars. If we're banking on this technology to support future robotic or human missions to the planet—and we are—there's no time like the present to begin testing it.

The thirty-three-pound MOXIE will be mounted underneath the front right side of the Mars 2020 rover. The process it uses is called solid oxide electrolysis, which is based on the fact that when heated certain ceramic oxides become oxide ion conductors. A series of ten membrane-electrode combinations, each consisting of a thin ceramic oxide membrane sandwiched between two electrically charged electrodes, are compressed and packaged into a thermally insulated, heated box. This box, coupled to a Mars atmospheric pump and the operational electronics, makes up MOXIE. It is complex to describe but relatively simple in operation.

Fig. 15.1. Principal investigator Michael Hecht with the MOXIE experiment, which is enclosed inside a Mars atmosphere simulator to left. Image from NASA/JPL-Caltech.

MOXIE is designed to create about eight grams of oxygen per hour—just a tiny amount, but enough to validate the technology. It is a subscale experiment, and if successful NASA plans to follow it with a much larger device, capable of producing and storing oxygen for future use by a Mars Ascent Vehicle for a soil sample return, and later for human spaceflight needs. And if plans like those of Elon Musk come to fruition—recall that his company, SpaceX, is building a huge Mars rocket called the BFR (for Big Falcon Rocket)—it will need lots of oxygen on the red planet to make routine runs there. Musk plans to begin transporting people to Mars in the 2020s and to eventually enable the creation of an entire city there. Oxygen and other resources, such as water from melted ice, hydrogen stripped from the water, and glass and metals from the soil, will all be important toward settling Mars.

The MOXIE experiment represents the first time such a complete system has been designed to operate autonomously, and to withstand the rigors and stresses of launch, interplanetary transit, and landing on Mars—no small task. But before humans can spend any appreciable time on Mars, we need to know that this kind of technology is workable there, and then follow those experiments up with full-scale, industrial-strength, pre-positioned robotic stations to create the supplies we'll need.

NASA's plans are to follow MOXIE with a hundred-times larger unit that will be nuclear powered and capable of processing up to two kilograms of oxygen per hour and storing the gas for later use. An open question is whether NASA's astronauts will end up sipping Martian martinis in Musk's already established MarsTropolis when they arrive, or if they will end up being the first to land on the planet. We should know within a decade—check with me then.

Chapter 16

BARNSTORMING VENUS: PART 1

Venus has never been the darling of the solar system when it comes to planetary exploration. It is true that some of the earliest robotic spacecraft were sent to the planet, and the Soviet Union made great strides when it landed its hardy Venera landers there a few years later, but since then the planet has only been visited intermittently. It's not for any lack of scientific interest, it's just that Venus is the backwoods, *The Hills Have Eyes* trailer park of the solar system.

Once upon a time—say, about the same time that people thought Mars might be crisscrossed with canals built by an advanced civilization—it was thought that Venus might be the primeval planet. With boundless optimism, learned academics speculated that, since Venus is closer to the sun, and therefore warmer than the Earth, it followed that the planet might be one giant marshy swamp, perhaps even inhabited by primitive life forms. Some less-pedigreed souls even suggested there might be dinosaurs stomping about. This was a steamy, romantic vision of the friendly solar system we would have loved to find. It's a damn shame it didn't work out that way.

Venus was instead revealed to be a true hellhole of a planet. With an atmosphere of about 96 percent carbon dioxide, and its proximity to the sun, the planet is the poster child for runaway global warming, with temperatures of at least a balmy 864 degrees Fahrenheit. The atmospheric pressure is nothing to sniff at either (quite literally, since the air is suffused with sulfur dioxide), equivalent to that about two-thirds of a mile under Earth's oceans, or about 90 times that at sea level. You might call the planet dense in more ways than one. No swamps or dinosaurs for our closest neighbor, but a sunbaked, dry, and un-cratered landscape that makes Death Valley look like a day at the beach.

Nevertheless, at the dawn of the space race, reaching Venus with unmanned probes was a top priority. The Soviet Union initiated efforts to reach the planet in 1961, trying first for flybys, then impactors, and then later for orbiters and landers. It would take eleven tries and six years to attain success.

The US program fared better—in mid-1962, Mariner 2 successfully flew past Venus after Mariner 1 failed during launch. This was an endorsement of NASA's habit of flying robotic spacecraft in pairs when possible, allowing for one to fail but hopefully not both (the Soviets had also tried this, but inconsistently). Mariner 2 was a lightly modified Ranger lunar spacecraft that had no cameras and was capable of basic measurements of radiation and magnetic fields and micrometeoroid impacts. It also measured solar wind and a number of small solar flares during its mission. Due to the small lifting capacity of the rocket used to loft Mariner 2, the portion of the spacecraft relegated to scientific instruments had to be kept under forty pounds.

The United States followed this in 1967 with Mariner 5, the updated Mariner design that was similar to those being flown to Mars. It was originally built as part of the Mariner 3 and 4 series for a possible Mars launch, but with the success of Mariner 4 was repurposed for Venus through the addition of more thermal insulation and the removal of the TV camera—since the planet was permanently veiled with a dense layer of clouds, rendering the surface invisible, a camera was not thought to be needed. Had they included one, Venus would have looked like a smog-yellow billiard ball with some streaks across it. Mariner 5 was also a flyby mission, and the spacecraft is still in a long orbit around the sun, dead and slowly baking in the harsh light. Additionally, because it was a single, surplus Mariner, the spacecraft flew alone. But fortune smiled on the mission, which was successful.

About the same time as Mariner 5, the Russians succeeded with their much more ambitious Venera 4. It also consisted of a flyby spacecraft, but with a lander attached that disengaged from the main spacecraft before the encounter with the planet. The lander transmitted data from the planet's atmosphere during its descent, confirming that the atmosphere was overwhelmingly carbon dioxide, with a small amount of nitrogen, traces of oxygen, and a tiny amount of water vapor. The flyby portion of the spacecraft was eleven feet high, and the solar panels had a deployed span of about thirteen feet.

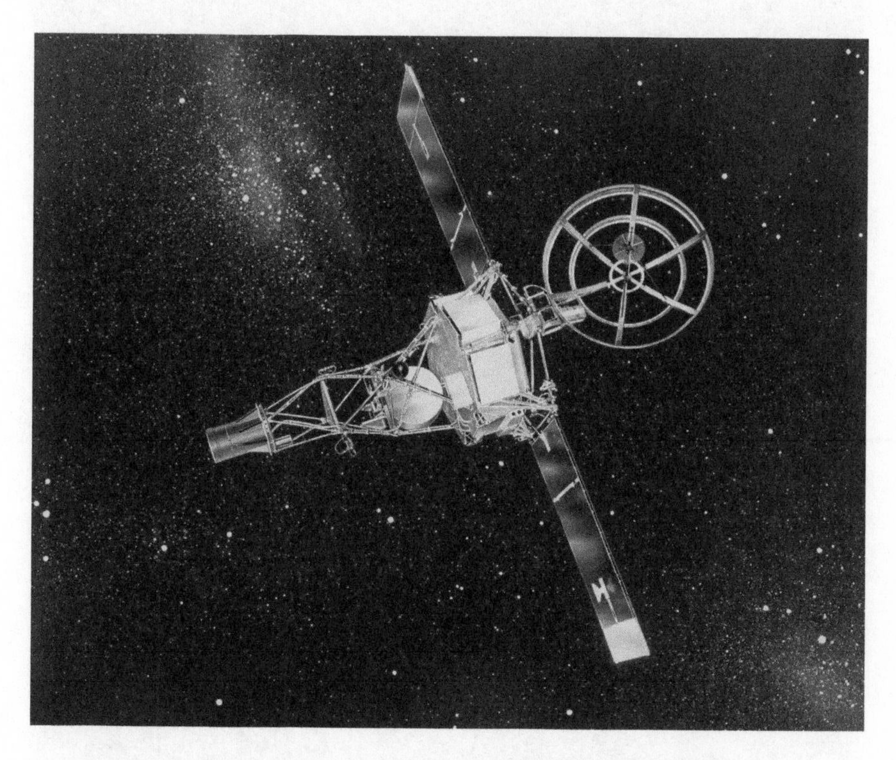

Fig. 16.1. Mariner 2, a repurposed Ranger design, accomplished the first successful flyby of Venus. Image from NASA/JPL-Caltech.

The landing stage was a metal ball three feet in diameter, with a gas-cooled heat shield. It was pressurized, a unique design that was characteristic of Soviet robotic explorers—they were almost like miniaturized crewed spacecraft. In this case, however, it was pumped up to over twenty-five times Earth's atmosphere at sea level, designed to withstand the crushing pressures thought to be endemic to the Venusian atmosphere.

Fig. 16.2. The USSR's first success at Venus, Venera 4. It was larger and more ambitious in scope than Mariner 2 or 5. Image from Wikipedia/Rave.

While it was considered unlikely, it was still thought there might be oceans on Venus. The Soviet engineers designed a very clever but simple arrangement to deploy the landers' antennas—sugar was cured into the seam of the hatch covering the antenna, and if the lander had come to rest in a liquid, the sugar would dissolve, releasing the antennae. Brilliant. Its burn-proof parachutes were tested up to a temperature of about 850 degrees.

During its ninety-minute trip through the dense Venusian atmosphere, the heat shield experienced temperatures up to 11,000 degrees Fahrenheit, and at thirty-two miles in altitude the parachute was deployed. Data on atmospheric pressure, temperature, and gas composition were transmitted back to Earth. At the point at which the parachute was released, the air temperature was only about 91 degrees, about the same as on that sunny day in Pasadena when Curiosity was about to land on Mars. But by the time the Venera drop probe stopped functioning, temperatures had reached 540 degrees and pressure was twenty-two times Earth normal—they had clearly not yet gotten to the even hotter ground level when Venera 4 packed it in. Nonetheless, it was the first set of direct measurements of another planet's atmosphere and its chemical composition, temperature, and pressure. But the Soviet Union wasn't done with Venus yet, not by a long shot.

Four attempts later, they succeed in reaching the surface with a functioning spacecraft. The intervening flights had been a mixed bag of launch and landing failures, each a lesson for the next attempt. In August 1970, Venera 7 plummeted to the planet's surface, making a successful, though lopsided, landing. It transmitted data for twenty-three minutes before dying, and was the first radio transmission to originate from another planetary surface. In July 1972, Venera 8 entered the planet's atmosphere at about 25,600 miles per hour, using a heat shield to slow in the dense air to a speed of about 560 miles per hour before opening its parachute at an altitude of thirty-seven miles and landing upright. Its data generally matched that of Venera 7. The temperature was over 870 degrees, with greatly diminished light due to heavy cloud cover, and an onboard spectrometer indicated that the nearby surface was granitic in composition. Venera 8 operated for just over an hour.

A few months later, in November, NASA succeeded in flying another Mariner past Venus, Mariner 10. But this was a flyby of convenience; the primary target for the spacecraft was sunny Mercury, which it reached in September 1974. While experiencing nearly five times the solar radiation of that near Earth, the probe managed to image both Venus and Mercury with its TV camera and returned copious amounts of data.

In 1975, the Soviets finally scored big. Never mind that they had been beaten to a manned lunar landing by the Americans—they finally managed to land a sophisticated, advanced machine on the surface of Venus, the truest of no-man's lands. Venera 9 consisted of an orbiter and a lander, and held the additional distinction of being the first spacecraft to go into orbit around the planet.

The spacecraft was a behemoth, weighing almost 11,000 pounds—truly appropriate for a mission to Dante's seventh circle of hell. The ability to launch this heavier machine was due to the successful development of the newest Russian booster, the Proton, which is still in use. While the orbital component was primarily a relay station for the lander, it performed a range of experiments as it circled the planet, transmitting for months. But the lander was the real prize. After its high-speed entry, a series of parachutes delivered it to low altitudes, where it was bravely released about thirty miles above the surface, using an innovative aerodynamic shield to continue its descent through the scorching air—the atmosphere was now known to be so dense at this level that no parachutes were required. It plopped onto the rocky surface on its donut-shaped metal landing cushion at about fifteen miles per hour.

Time was of the essence, as the Soviets had learned that machines on Venus did not last long. Within two minutes, instrumentation had been activated and was returning data. Cooling was achieved via circulating fluids, and soon the first image ever transmitted from the surface of another planet was sent back to Earth—a desolate, flat, rocky plain that, while fascinating, was truly uninviting. This was especially true when the chemical analysis of the atmosphere was returned—bromine, iodine, hydrochloric and hydrofluoric acid were all part of the toxic brew—way worse than LA smog. The local temperature that day was 905 degrees Fahrenheit, and the vista was about equivalent in illumination to a cloudy day on Earth. No shadows were seen on the overcast surface, and the toxic air was clear. There appeared to be little weathering on the nearby rocks. While it might look like parts of the South Dakota Badlands, it was far less inviting.

Then, fifty-three minutes later, it was over. The lander succumbed to the awful heat, and, like its brethren, remains a scorched hulk to this day.

After a series of follow-on missions that were essentially similar in design—the Soviets had a de-facto assembly line going for their rugged Venus spacecraft—two remarkably new and evolved probes flew in 1981. Veneras 13 and 14 were a bit lighter than their predecessors, but far more capable. Both landed in March 1982, following the successful landing sequence of Venera 9 and its descendants. They landed about 590 miles apart, in a region now known as Phoebe Regio, and immediately got to work. The innovations on these landers were rather remarkable. Spring-loaded arms slammed down to measure the compressibility of the surface (Venera 14, by serendipity, slammed its sensor directly onto a lens cover that had popped off the lander upon touchdown, measuring its compressibility instead). Then, a screw-tipped arm pivoted down to obtain a surface sample, which was then deposited into a sealed chamber. There, an X-ray fluorescence spectrometer went to work, analyzing the bits of rock and soil ingested. Venera 13's sample was a type of rock known as weakly differentiated melanocratic alkaline gabbroids (no, it doesn't require medication), a type of volcanic basalt, and Venera 14's sample proved to be similar to oceanic basalts, also a type of volcanic rock.

Venera 13 functioned for over two hours, and Venera 14 for about fifty-seven minutes. Onboard microphones recorded the sounds of

Venus during that time, which sound like white noise, or static, but were nonetheless the first recorded sounds from another world.

Fig. 16.3. An image of Venus from the short-lived Venera 14. The surface was anything but hospitable, and a far cry from what had been imagined just fifty years earlier. Image from LP Archive.

The Soviets were on a roll with Venus, and they scored a series of successful missions thereafter. Orbiters mapped the surface with radar in the mid-1980s, and more landers based on the capable Venera design followed, renamed Vega. Of particular interest, the Vega missions used an atmospheric balloon probe made of Teflon, released at about thirty-seven miles' altitude and immediately inflated with helium. The air was cooler here, and the balloons from each craft drifted at a stable altitude of about thirty-three miles, at a speed of about 150 miles per hour. The balloons continued to send back data for a few days each before failure. Air pressure at this altitude was about half of Earth's at sea level, and the temperature was about one hundred degrees Fahrenheit. Using balloons on another world was another Soviet first.

The remainder of the Vega spacecraft, the parts that stayed in space, were not orbiters, but performed flybys instead. Each utilized a gravity assist from Venus and were redirected to image Comet Halley, then making a close approach to the sun. On March 6, Vega 1 returned more than five hundred images of the comet's nucleus, showing jets of vapor erupting. Vega 2 performed the same feat three days later, returning about eight hundred images.

Not to be left entirely out of the Venus game, NASA flew two

Pioneer probes to the planet in late 1978. The first was an orbiter that looped the planet for fourteen years, returning data until 1992. The second component, Pioneer Venus 2, was more complex. The eight-foot-wide main body of the spacecraft glided into the Venusian atmosphere to return measurements of the upper layers. The spacecraft was unshielded—it simply returned data until it overheated and broke up. However, en route, it had released four more small probes designed to descend into the deeper atmosphere; only one—larger than the others and more heavily instrumented—had a parachute to slow its descent. The probes were not designed to survive impact, merely to send rapid streams of data back as they passed deeper and deeper into Venus's hot atmosphere. Surprisingly, one of the small probes did manage to continue transmitting for almost an hour after hitting the surface.

In mid-1989, the American Magellan orbiter reached Venus, remaining active in orbit around the planet for over four years. In that time, it made a comprehensive and high-resolution radar map of the surface—radar is the only way to peer below the dense cloud layers. Magellan's map covered 98 percent of that world. In October 1994, Magellan entered the atmosphere and more experiments were conducted during its destruction.

Fig. 16.4. Magellan's first radar map of the surface of Venus. While geologically fascinating, it is even less inviting than Mars. Image from NASA.

The maps generated by the mission, the only complete mapping of the planet, showed a surface violently affected by volcanic activity. Long, flat plains of lava flow, fields of small lava domes, and large shield volcanoes are common. Some lava flows run for about 3,800 miles, and although Venus is riven by rift zones and topped by domes indicative of the upwelling of magma, there is no firm evidence of plate tectonics as is seen on the Earth. There does not appear to be much erosion of surface features either, and little windblown transport of dust or sand.

Since that time, the United States, the European Space Agency, and the Japanese Space Agency have flown additional missions past or around Venus, but none so audacious as those of the Soviet Union—and to date, nobody else has dared to land on Venus's broiling surface. But there have been some interesting ideas . . .

CHAPTER 17

BARNSTORMING VENUS: PART 2

There were other Venus programs brewing in the United States in the 1960s, and one of them would blend humans and robotic probes in space. At the time, the use of robots in space was still considered to be an ongoing experiment, despite the successful flybys of Venus and Mars earlier in the decade. Early computers and associated hardware were still considered unreliable, and the science return was thought to be small in comparison to what might be accomplished with humans on hand. And while JPL's engineers did fine work remotely controlling these robots from their home base in California, what glories might be achieved if the puppet masters were closer to the machines they controlled? With shorter radio-signal travel times, and the ability to respond to challenges and opportunities on-site immediately—on-site in this case being near Venus—it was thought that much more might be achieved. Never mind that the cost would be exponentially higher to accomplish such a mission. This human-centric thinking was the impetus for a proposed program to send American astronauts looping past Venus with their robotic associates doing the heavy lifting on the forbidding planet below.

As detailed in my book *Amazing Stories of the Space Age*, there were a variety of schemes put forth from the late 1950s onward to fly astronauts past Venus and Mars.[1] It's a long story of wishful thinking, big ideas, and scuttled dreams. In reality, it would have been a long trip in cramped quarters with a high likelihood of the humans aboard being exposed to deadly levels of radiation, especially if there had been any solar events of even medium caliber—none of the designs put forth for these missions incorporated what would today be considered a safe level of radiation protection. But these dangers were not well understood at the time, and the notion of a manned flyby of other planets seemed like a logical extension of the Apollo lunar landing program.

One of these schemes in particular deserves retelling, for the idea of suiting up a trio of astronauts and sending them off on a looping flight past Venus, where they would drop their mechanical stand-ins to suffer the depredations of the Venusian environment as the humans joysticked the robotic operations as their ship swung past the planet, is just too good to ignore.

By the mid-1960s, Apollo was running full-bore throughout NASA's empire—even the robotically oriented JPL was flying the Ranger, Lunar Orbiter, and Surveyor programs in support of the manned lunar landing effort. But at places like NASA's Marshall Space Flight Center in Alabama, where Wernher von Braun and his team were masterminding the creation of the Saturn V rocket, the conveyor belt of money coming from federal coffers was slowing down—by 1966 they would begin to feel the real pinch. They had developed the Saturn V moon rocket and were completing enough of them for Apollos 8 through 20 to make their lunar journeys, but unless something came along to incentivize the government to build more of these rockets, Marshall would soon be losing a lot of their employees and expertise, an eventuality that von Braun abhorred. The escalating war in Vietnam was also increasingly expensive, and the pressure was on to cut NASA's funding.

Von Braun had long fostered more ambitious plans than just landing on the moon. He had been an advocate of human missions to Mars since the 1940s, and he felt that the Apollo hardware was a strong potential base upon which to build other exploration programs. There were a number of possibilities—a lunar base was considered, as were various Earth orbital stations based on Apollo hardware. As mentioned, Mars was also a prime target, and Venus was also considered. Any of these goals would need big rockets to do the heavy lifting, and both the Saturn IB and the Saturn V were perfect for such chores.

With von Braun's support, and that of other members of NASA

senior management, the Apollo Extension Series (AES) office had been founded in Washington to look for further uses for the magnificent hardware that was being designed for the lunar landings—to extend its usefulness beyond Kennedy's goal of simply landing a man on the moon. This evolved into the Apollo Applications Program (AAP) in 1968. A number of program proposals came out of these offices, but the only one ultimately enacted was Skylab, America's first space station, in 1973.

One of the more unusual ideas the fertile minds at the AES office had been toying with was planetary flybys. Landing on the moon would be hard with 1960s technology, and landing on another *planet* even more so given the need to transport exotic landers the vast distances that separate the inner planets.

These were based on a 1956 paper by an Italian engineer, Gaetano Crocco, which, while initially suggesting a robotic mission, in its final iterations would lead to a plan for a manned flight that would follow a long, looping trajectory past both Mars and Venus—a flyby with no landing, with far less mass and complication required.

In 1962, further studies of Crocco's ideas had been commissioned—based both on NASA's current understanding of its upcoming Apollo hardware and also on new designs—from a trio of aerospace contractors. The reasoning was simple, and it was expressed succinctly by Harry Ruppe, an engineer at Marshall involved in the studies. Ruppe thought that a logical progression beyond the moon was imperative: "From the lunar landing in this decade to a possible planetary landing in the early or middle 1980s is [a span of] 10 to 15 years. Without a major new undertaking, public support will decline. But by planning a manned planetary [flyby] mission in this period . . . the United States will stay in the game."[2] The most ambitious of these plans would fly past both planets and take about two years. At the time, both the Apollo spacecraft and the Saturn rocket were under development, and while using that hardware would be optimal, the planetary flyby spacecraft's overall design was in flux.

None of the 1962 studies resulted in a commitment from NASA's masters in Washington. One downside was that each contractor suggested new spacecraft designs, optimized for interplanetary travel, which would have resulted in vastly increased expenses beyond what was already being spent on the Apollo spacecraft hardware. Consequently, a follow-on study, constrained to using Apollo hardware, was commissioned with NASA planning contractor Bellcomm for delivery in 1967. This hardware would consist of the Apollo capsule, its propulsion unit, and the Saturn's S-IVB stage, the moon rocket's upper stage, which would be used as a habitat for the long interplanetary voyage. The flight past Venus would use a "free return" trajectory, meaning that, other than corrective maneuvering, once the spacecraft set off from Earth it would swing past Venus with little additional intervention—no additional rocket stages would be required there since they would not be braking into orbit around that world—and they would then head back to earth for reentry many months later.

As specified in the report's introduction:

> This report first describes a specific planetary reconnaissance program, and a probe complement is planned to accomplish the scientific and engineering objectives set forth. The various aspects of Venusian atmospheric entry and the constraints imposed on time line sequencing and probe design are considered.[3]

The 1967 report concentrates on a flyby of Venus in 1973–1974 with three crewmen aboard, along with a variety of robotic probes to be dispatched to the planet during the flyby. Given the availability of the two Apollo launch facilities at the Cape, a dual mission with two matching sets of spacecraft was also considered. During the interminably long voyage, telescopic observations of planets and nearby Near-Earth Objects (NEOs)—aka asteroids—would be accomplished via a forty-inch diameter telescope carried along—itself a major accomplishment, as at that point no telescopes of any size had been flown in space. There were many advantages to using a large telescope in space—not only would the spacecraft be closer to many of these targets during the flight than they would be on Earth, but without the distorting qualities of Earth's atmosphere the images and observational data would have been spectacular. Key observational targets also included the sun and Mercury, which would be about 31 million miles distant at their closest. The trip out would have taken 123 days, and the total trek 273 days—a long time for a crew living in small quarters.

The key objectives would be those accomplished during the Venus flyby phase. As the spacecraft neared the planet, various probes would be dispatched to the planet. This single mission would have accomplished most of our Venus exploration goals in one grand journey of under a year, as opposed to the individual robotic missions we eventually flew in the decades of the 1960s and 1970s.

Longer multi-pass journeys were also discussed in the study for 1977–1978 launch windows, but the single flyby 1973–1974 mission was the primary point of reference.

The various robotic probes that would be sent Venus during the flyby were:

Drop Sonde/Atmospheric Probe (DSAP): A capsule that would make a steep entry into Venus's atmosphere, then once it was sufficiently slowed would release a smaller probe that would make atmospheric measurements as it descended toward the surface.

Orbiter: An orbiting probe that would be released to brake into Venus's orbit using small rocket engines and would conduct observations from an orbit around the planet long after the astronauts performed their all-too-brief multi-hour flyby.

Meteorological Probes: Designed along the lines of high-altitude weather balloons and sensor devices used on Earth, these probes would enter the atmosphere packed into capsules and be released at various altitudes to drift among the high-altitude clouds of Venus, collecting data about these atmospheric regions for several weeks.

Photo/R. F. Probes: These probes would be modeled after the Ranger spacecraft that had reconnoitered the moon, although they would be enclosed in aerodynamic shells. The hope was that once they cleared the densest cloud belts surrounding Venus, they would be capable of imaging the surface as they crossed the last few miles to the ground. They were not designed to survive impact with the surface.

Lander Probes: These would be manufactured in two sizes—the larger one would, once on the surface, perform photographic surveys, soil analysis, and weather measurements. Smaller versions would simply monitor weather conditions on the planet for as long as they lasted.

High-Altitude Buoyant Venus Device (BVD): This larger balloon would conduct biological experiments to seek life in the surprisingly cool and benign upper atmospheric layers of the planet. It was hoped that they might survive in the cooler environment high above the hellish surface for up to six months.

Near Surface Floater (NSF): This balloon would operate at one thousand to five thousand feet above the Venusian surface, beneath the densest cloud cover, and would be designed to descend to hundreds of feet if practical. Imaging and weather measurements would be the key priority. It would be anchored to the surface to allow elevation changes via winching in or out the attaching tether. A small surface sampler and attached TV camera was also considered for deployment from the floater.

The Drop Sonde probes would be the first to be released, about sixteen hours before the flyby, and the landers would be dispatched closer to the planet to allow for direct control of their descent—it was considered impossible to control them from Earth because of the long delay in radio signals from the landers back to our planet, well over six minutes, depending on the distance at the time. The orbiter would be released at the manned spacecraft's furthest point from Earth on the far side of Venus. At least some of the probes were expected to be powered by long-lived nuclear fuel sources.

The numbers of probes and the order of dispatching them depended on which mission profile was followed, but in any scenario the crew would have been extremely busy managing these machines—first sending them on their way to Venus as they approached, then overseeing their first observations as the crew looped the planet.

As the crew sped away from the planet, they would head back to Earth at an ultimate return speed of over 30,000 miles per hour, a good 5,000 mph faster than the Apollo missions' returns from the moon. Additional work would have been needed to test the Apollo capsule's heat shield for the mission—it is still a challenge today, even with modern materials, to build heat shields that are reliable in this temperature range, well above 5,000 degrees Fahrenheit in the case of a return from Venus.

It should be noted that by the time the earliest of these manned missions would have flown, the Soviet Veneras were already reconnoitering the atmosphere and surface of Venus robotically. It could have been embarrassing had the Apollo-based hardware been preparing for a 1973 launch when Venera 8 landed in 1972, or if the 1977 launch window was being prepared when Veneras 9 and 10 landed successfully in mid-1975.

In any case, the plans were too ambitious for Congress, which had been wielding their budgetary axe over the still-to-fly Apollo program for a year already, and would make further cuts in 1969 and beyond. By 1973, the last Saturn rockets—which had been intended for the now-canceled missions of Apollo 18, 19, and 20—were headed toward museums, and Skylab would wrap up its last crew visitation in the beginning of 1974. The Apollo hardware would be used for one more mission, the anticlimactic docking of an Apollo capsule with a Russian Soyuz in 1975, and that was the end of the program.

In the end, a crewed flyby of Venus was not to be. It was overly ambitious for the hardware of the time, and the radiation hazard was not completely understood or worked out. The small confines of the docked capsule and repurposed S-IVB stage would have taxed even the hardiest crews over the course of almost a year, and the arguments for controlling the robotic probes from a passing spacecraft were flimsy at best. Nevertheless, had the project been undertaken, and the crew returned safely to Earth (without any lasting effects from space radiation), it could have been glorious.

Chapter 18

FLASH FORWARD: THE TICK-TOCK ROVER

The future exploration of Venus will probably be accomplished via the use of a rover, as, in an endlessly fascinating, varied landscape, mobile explorations are much more productive than static landers, which are destined to examine just one spot. But what kind of autonomously driving machine could survive such an extreme environment? Even with efficient, modern refrigeration systems, and new, temperature-tolerant materials, keeping any type of machine working for long stretches of time in near-900-degree Fahrenheit heat is a huge challenge.

Steampunk to the rescue!

A new Venus rover called AREE, for the Automation Rover for Extreme Environments, is currently being studied at JPL. Referred to informally as a "clockwork rover," it looks like a steampunk coffee table with tank treads on either side. Its designer, Jonathan Sauder, said that he was inspired by mechanical computers, such as those in use in the first half of the twentieth century. These used cranks, levers and cogs, as opposed to electronics, to compute figures. The use of such classic analog mechanical devices could succeed on Venus, where electronics would simply melt.

"Venus is too inhospitable for the kind of complex control systems you have on a Mars rover," Sauder said. "But with a fully mechanical rover, you might be able to survive as long as a year."[1]

Another engineer on the project, Evan Hilgemann, added, "When you think of something as extreme as Venus, you want to think really *out there*. . . . It's an environment we don't know much about beyond what we've seen in Soviet-era images."[2]

Given that even the heartiest Soviet probes only lasted hours, this new mechanical approach may be the only way to accomplish a long-term exploration of Venus's surface. The tank treads were inspired by First World War "mother tanks," which had treads that wrapped all the way around each side, from top to bottom. The idea was that they would be able to traverse trenches and shell holes—terrain not too dissimilar from what is expected on Venus.

Another challenge in high-temperature environments is communications—again, like the other electronics, the components would melt. The AREE will use a system of radar reflectors on the top of the rover, covered by a set of mechanical shutters. The shutters would rotate to expose or cover a part of the radar reflector, resulting in a very low-speed mechanically driven code, much like Morse code. Orbiters, using radar, would pulse their signal at the rover, and the radio signal return would be coded mechanically, not unlike the way ships at sea use flashing lights to communicate.

The machine's motive power will be generated by wind turbines located within its open-frame chassis. The turbines do not charge batteries, which would also malfunction and melt, but will be used to directly drive the rover, and also to store energy in springs that can be harnessed later for driving the communication shutters. While the plan is for AREE to have backup solar panels that would be armored against the heat, much of the investigative work will be done using analog mechanical instruments.

The AREE rover is even being considered for Mercury's sunlit side, and for extremely high radiation environments such as the moons of Jupiter and Saturn, where cold and radiation conspire to kill conventional electronics and machinery.

Maybe contemporary fiction author Scott Westerfeld was on to something when he wrote, "The Internet is global and seemingly omniscient, while iPods and phones are all microscopic workings encased in plastic blobjects. Compare that to a steam engine, where you can watch the pistons move and feel the heat of its boilers. I think we miss that visceral appeal of the machine."[3] He may be proved correct when we marvel at the discoveries resulting from AREE's drive across Venus, as it unravels the mysteries of one of the least explored inner planets.

Fig. 18.1. JPL's Automation Rover for Extreme Environments, or AREE, has wind-powered treads that wrap around its sides, and it communicates via mechanical shutters and Morse code. No melt-prone computer chips need apply—just wind it up and go. Image from NASA/JPL-Caltech.

The picture that inflamed the reporters: the very first image from the surface of Mars, a shot of Viking 1's footpad. The engineers wanted to know that their machine had landed safely—a horizon shot would come soon thereafter. *Image courtesy of NASA/VMMEPP.*

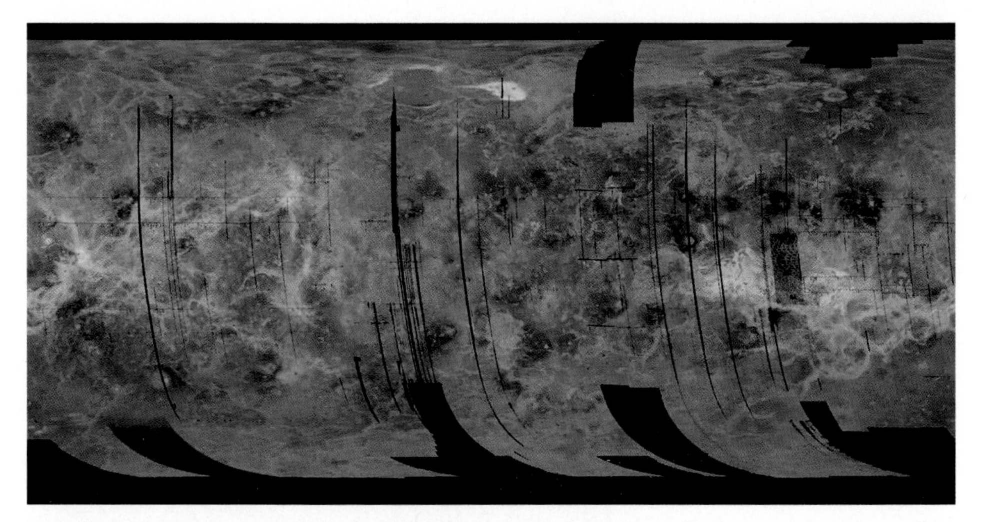

Magellan's first radar map of the surface of Venus. While geologically fascinating, it is even less inviting than Mars. *Image courtesy of NASA.*

A Voyager control room, 1970s, looking very state of the art for the time. *Image courtesy of NASA/JPL-Caltech.*

A volcano erupting on Io. This image has been enhanced to bring out detail in the eruption. You can imagine Linda Moribito's surprise when she realized what it was. *Image courtesy of NASA/JPL-Caltech.*

An exaggerated-color image of Titan's limb as seen by Voyager 1. You can understand why people got so excited at times like this. *Image courtesy of NASA/JPL-Caltech.*

Uranus, the planet with a name you can't pronounce without engendering giggles from someone in the room. It was as smooth and featureless as a billiard ball. *Image courtesy of NASA/JPL-Caltech.*

Magnificent Neptune, with far more of interest in the upper atmosphere for planetary scientists than Uranus. *Image courtesy of NASA/JPL-Caltech.*

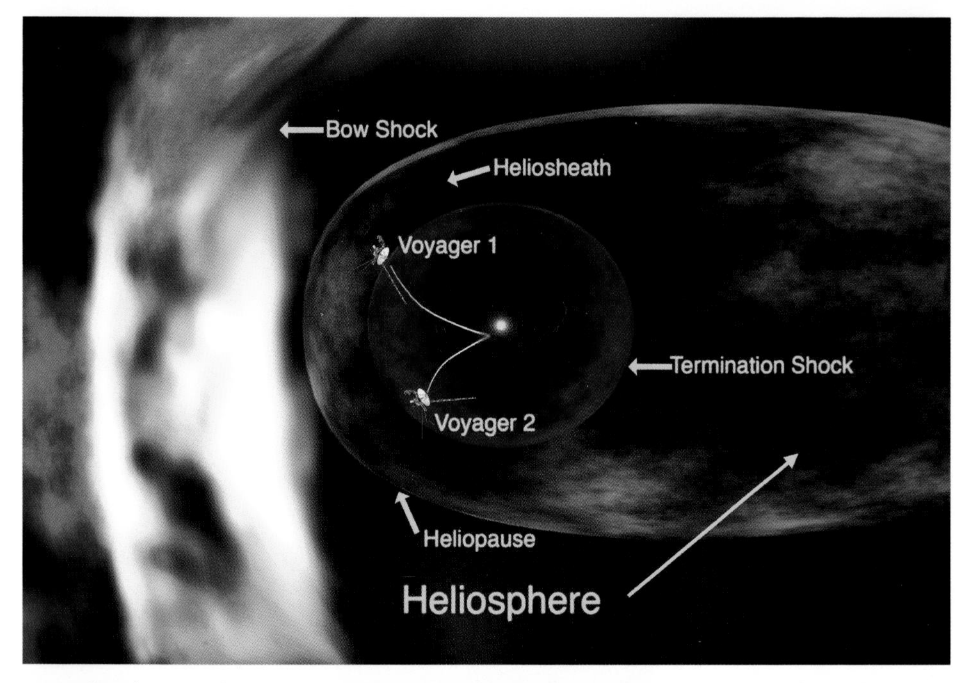

A graphic showing the approximate moment that Voyager 1 *(top center)* passed through the termination shock and entered the heliosheath. *Image courtesy of NASA/JPL-Caltech.*

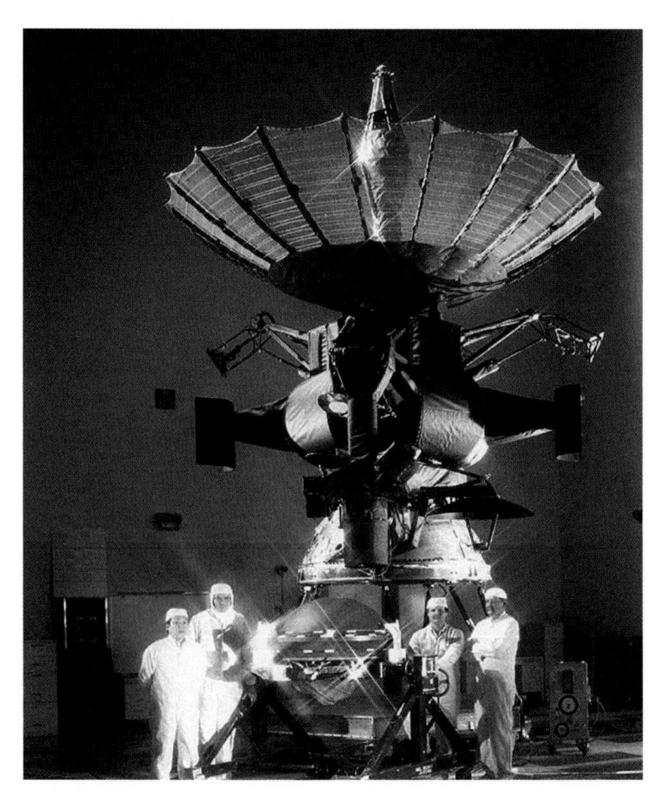

Galileo seen with engineers in 1983, showing how the antenna was *supposed* to work. In practice, it barely started to unfold before becoming snagged. *Image courtesy of NASA.*

Cassini's portrait of Jupiter, taken in December of 2009 en route to Saturn. *Image courtesy of NASA.*

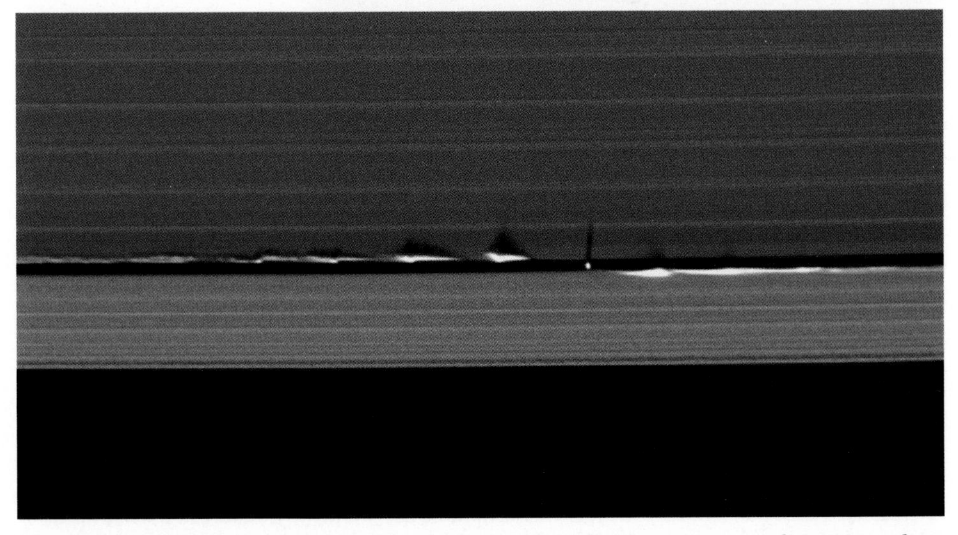

The tiny moon Daphnis, seen as a bright spot just to the right of center, causes disruptions along the edges of the surrounding rings, seen here as white "foam" with shadows cast inward. *Image courtesy of NASA/JPL-Caltech/SSI.*

The beginning of an ongoing storm caused by Saturn's recurring Great White Spot in 2010. After it's gone on for a year or more, a series of interconnected white spots can cover the better part of a hemisphere. *Image courtesy of NASA/JPL-Caltech/SSI.*

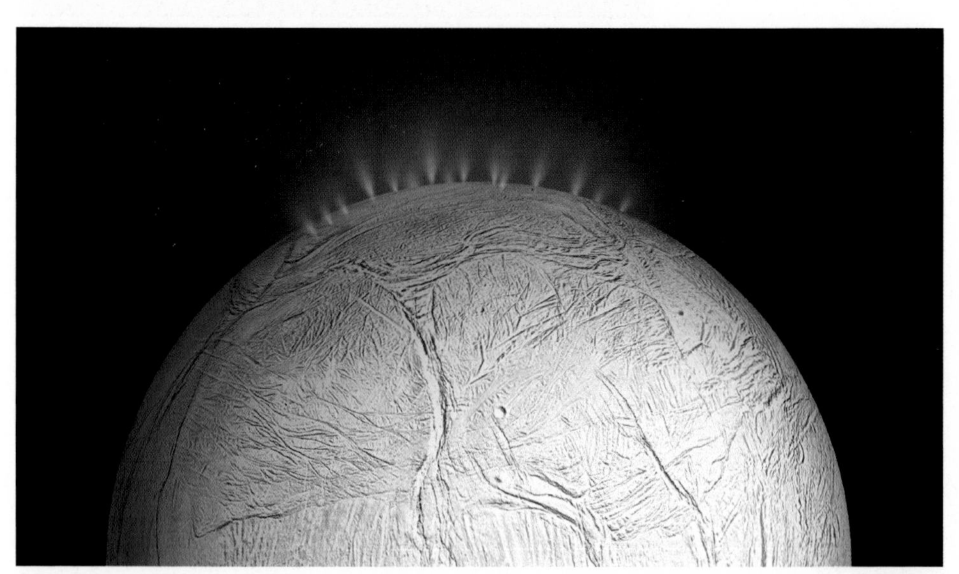

Enceladus: a moon of mystery. Geyser-like eruptions are frequently observed jetting up from the surface. These are eruptions of water from below the icy surface, caused by tidal flexing due to the intense gravitational squeezing and warming of the moon. *Image courtesy of NASA/JPL-Caltech/SSI.*

A view of Titan's surface from the Huygens probe, processed for contrast. If Pluto were still a planet, Titan could almost be considered one as well. *Image courtesy of ESA/NASA/JPL/University of Arizona.*

Artist's impression of the end of the Cassini mission, as the spacecraft begins to heat up in Saturn's atmosphere. To the sentimentalists in the room, and there were a number of us, it felt like a valiant moment. *Image courtesy of NASA/JPL-Caltech.*

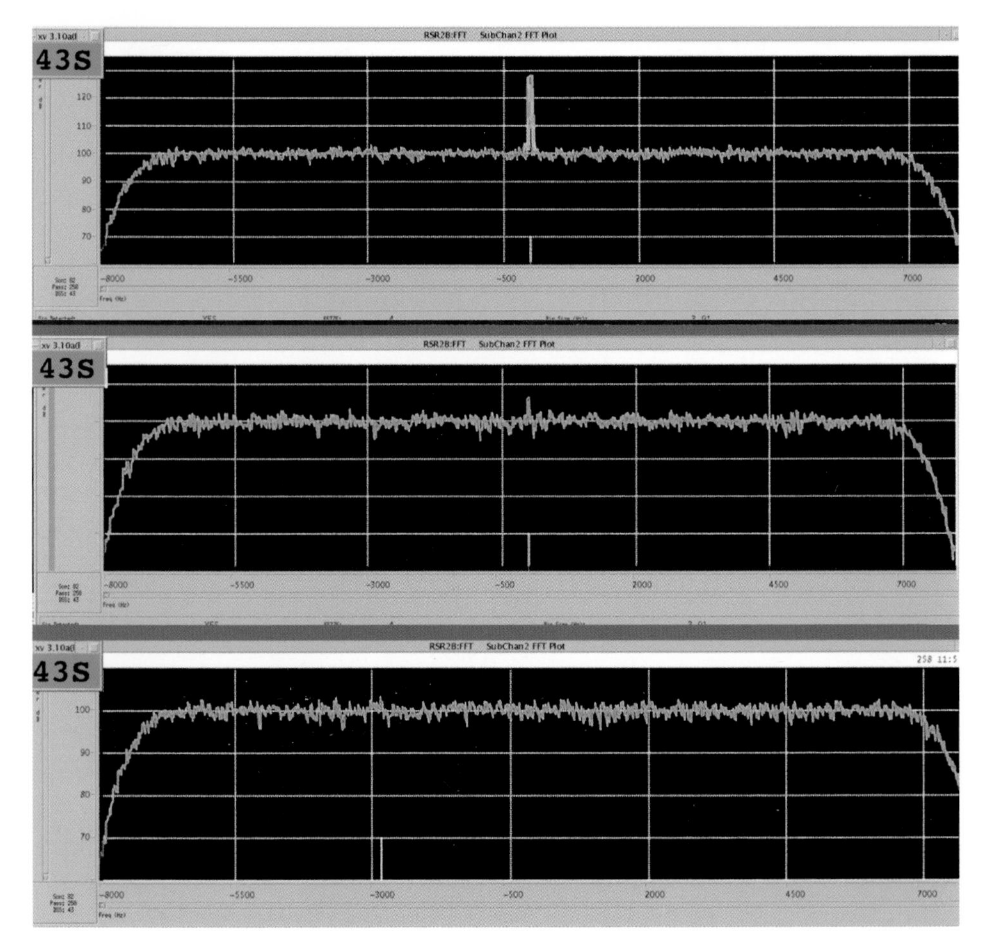

The readout of the S-band radio signal as Cassini met its end—the signal strength of the radio is represented by the spike at the top center of the squiggly line. The image at the top is a few minutes before the loss of signal; at the middle, the signal is fading; and at the bottom, it is gone, as was Cassini. *Image courtesy of NASA.*

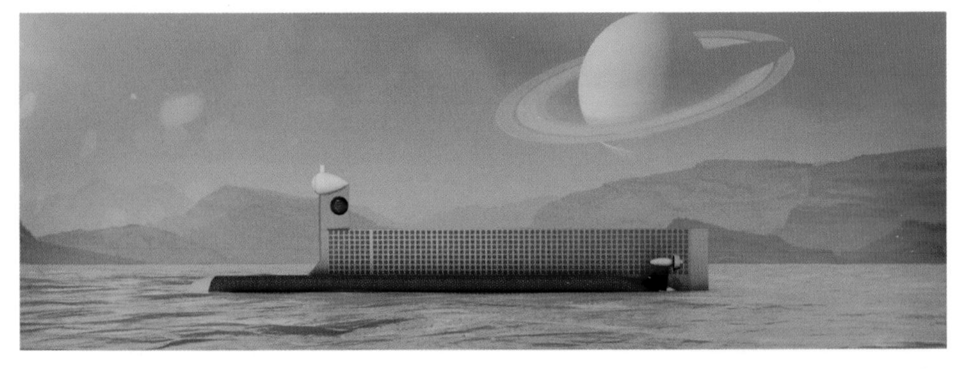

One of the designs for a future submersible that would explore the frigid seas of Titan. The submersible would be highly autonomous and would be forced to surface to communicate with Earth. The fin on the back is an antenna for sending radio transmissions to an orbiter or directly to Earth. *Image courtesy of NASA/NIAC.*

CHAPTER 19

THE CENTER OF THE UNIVERSE: PART 3

By 8:00 p.m. the assembled reporters, JPL employees, and various NASA officials that were not in mission control were gathered in Von Karman Auditorium to witness the last couple of hours before Curiosity's touchdown on Mars. Von Karman is a good-sized venue, and a few hundred seats were filled with reporters who were chatting each other up (doubtless comparing the penurious rates of pay for their articles) and enthusiasts of various stripes who were fortunate enough to get a press credential. The television crews were set up across a riser toward the back, and behind them were small workstations for the social media attendees. The venue was buzzing with excitement—something between breathless anticipation of imminent success and a deathwatch for a mission some feared would fail spectacularly. One group was sure to be disappointed tonight—which of the two would become clear in about two and a half hours.

I had claimed a seat mid-auditorium. Between eight and nine-thirty I wandered a bit, chatting up old friends and making new ones. The space news community is a fairly tight group—I think we fancy ourselves as something like the foreign correspondents who gather in various war zones around the world at hotel bars, awaiting the latest developments from the battlefront, except that we are infinitely safer and in this case would head home soon after the landing. The divisions among this cluster of journalists, if they can be called that, tend to fall along the lines of age—those of us who were alive during the space race, and those who came of age sometime during the shuttle program. There are other possible divisions by media outlet—major national and international publications and outlets versus smaller ones—but these distinctions are slowly vanishing as digital takes over the media landscape.

Many reporters and feature writers were joining the rest of us virtually, via NASA's webcast, from all over the world. This twenty-first-century media setup is a far cry from the days of Viking or Voyager, when one would actually have to attend in person to "get the story." These days, it's mostly a matter of preference, and for many professionals attending in person is a costly extravagance.

As the evening wore on, attention was divided between the computer-generated image of Mars from the perspective of Curiosity, looming ever larger on the screen in the front of the room, to increasingly excited side conversations about the mission and other space happenings. The entire crowd was beginning to feel a small fraction of the stress that the mission controllers were experiencing over in the SFOF . . . In just a couple of hours, it would be over, one way or the other. Either the grandest mission in the history of Mars exploration would be poised to embark in Gale Crater, or a few acres of red soil would bear the scars of mankind's greatest experiment in planetary exploration.

I looked up and saw the kilometers remaining to reach Mars rapidly diminishing, and the planet, already filling much of the screen, got one notch larger.

Chapter 20

A GRAND-ISH TOUR

Talk about your once-in-a-lifetime opportunity . . . how about a once-every-175 years' opportunity to whiz by the planets of the outer solar system? That's a bucket list item for sure. And when JPL planetary scientist Gary Flandro published his 1966 paper entitled "Fast Reconnaissance Missions to the Outer Solar System Utilizing Energy Derived from the Gravitational Field of Jupiter,"[1] some took notice . . . here was an opportunity to compress what could otherwise take thirty-plus years of travel time down to a bit more than a third of that.

What Flandro had noted was that the outer planets would be aligned in the late 1970s in such a way that a spacecraft flying past them could take advantage of the gravitation of each of those bodies to sling itself at ever higher speeds to the next one. This was not a linear alignment, mind you, as was presented in some popular artwork of the 1970s, but rather a curving arc with Jupiter in the lead and Neptune following at the rear, and with Saturn and Uranus in between. Flandro's realization centered on the fact that a spacecraft could use a spiral-like, curving trajectory to fly past each planet before exiting the solar system, which would be a bargain in time, energy, and, perhaps most importantly, money. Another such alignment would not come until the twenty-second century.

The possibility of a series of flybys of the outer planets first had first been considered at a 1965 meeting of the National Academy of Science's Space Science Board, held in Woods Hole, Massachusetts. The results of that meeting encouraged NASA to expand their goals to the planets, beyond what had already been done with Venus and was underway at Mars, which they also heartily endorsed. The focus was on a reconnaissance of the outer planets via flybys, or alternatively a more intensive study of only Jupiter, using orbiters and drop probes. The onus was on smaller, less expensive missions that could evade the budget-cutting axe that seemed to befall large, flagship missions.

Regardless of budget-axe fears, however, an interest soon formed around the notion of a what NASA called a "Grand Tour" of the outer planets based on Flandro's ideas—his 1966 paper, which he had been working on since 1964, underscored the viability of such a project—on a shorter timeline. Launch windows for this could be found between 1976 and 1980, and it was quietly pointed out that such a mission would, in the end, be cheaper to conduct than a bevy of individual missions to each planetary system, as originally considered by the Woods Hole meeting. As Don Hearth, then the director of NASA's Planetary Programs Office, said in 1988, "You've got to remember selling a new start is a bitch ... it's even worse today, but even then. It's almost as hard to sell a hundred-million-dollar project as it is a billion-dollar project. And a hell of a lot more work to sell two $100-million projects than one $200-million project."[2]

Flandro's design for a Grand Tour had been sold—at least to the chief of NASA's planetary programs. At about the same time, JPL began advocating that it be entrusted with the Grand Tour, while other proposals came in from other NASA field centers.

To arrive at a fundable program, NASA formed the Outer Planets Working Group in 1969. By this time, Apollo funding was winding down, and new projects were needed to keep the space agency viable—the Grand Tour looked like just such a project and had the added advantage of being high profile, resulting in great PR for NASA. It would be expensive, but a budgetary flyspeck compared to Apollo. As the field centers wrestled for ownership of the project, the focus shifted to two individual programs, one that would depart for Jupiter, Saturn, and Pluto in 1977 and another that would launch in 1979, bound for Jupiter, Uranus, and Neptune.

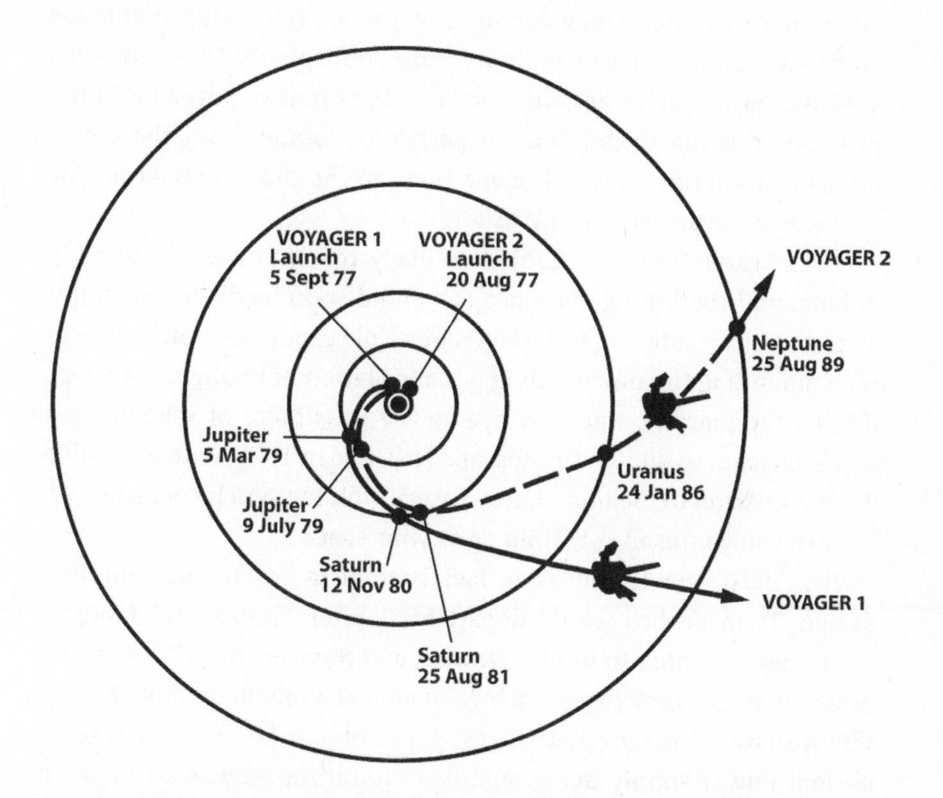

Fig. 20.1. The natural phenomenon that made the Grand Tour—and later Voyager—possible, an alignment of the planets that allowed for gravitational assist from world to world. Image from NASA/JPL-Caltech.

In 1969, incoming president Richard Nixon began cutting NASA's budget. Apollo's goals had been met, and space was no longer considered a proving ground for the United States against the Soviet Union. America had won that competition with the landing of Apollo 11, and other priorities took front-row seats, including the termination of the Vietnam War.

A schism resulted between groups, one favoring the fast-and-short missions focused on Jupiter, and the other favoring the Grand Tour, which would require spacecraft with longer lifespans but did not require the quick development of probes that could withstand entry into Jupiter's dense, high-pressure atmosphere. There emerged a third group, which was concerned that the funds required for either program, but the Grand Tour in particular, would delay the development of a large space telescope then on the drawing boards. The wicket was getting sticky in Washington.

The Grand Tour was quoted as likely to cost in excess of $700 million, and the Viking Mars project would soon be devouring funds at a prodigious rate—in fact, the Woods Hole group had come close to recommending the downscaling or cancellation of Viking due to cost. NASA headquarters was now eyeing the possibility of scheduling a single mission to Jupiter, Uranus, and Neptune in 1979, with a possible second mission to include Jupiter, Saturn, and Pluto to be determined.

Are you confused yet? Hold on to your seats.

By 1970, the Grand Tour had become a four-launch mission design. Twin probes would depart for Jupiter, Saturn, and Pluto in 1976, and two more to Jupiter, Uranus, and Neptune in 1979. The cost projections had now risen to a high of almost a billion dollars. Driving this cost was, among other things, development of a new, long-term nuclear power supply that would be required, as well as a computer with a primitive form of artificial intelligence that would be capable of taking care of itself during the long, extended mission through the solar system.

Then, into this fray, leapt another complication. It was called NERVA, and was a nuclear fission rocket engine that had been in development for years, with an eye toward human missions. NERVA represented a whole new approach to fast spaceflight, and it was considered critical to future NASA efforts beyond the short hops to the moon. Variations on the design had been tested on Earth, and the engine showed great promise. It was far more powerful than chemical engines of equivalent mass and could fire for much longer periods of time. A Democratic senator named Clinton Anderson, an important figure in spaceflight at the time, was steadfastly against anything that would divert funding from his nuclear rocket. And while he felt that

NERVA was perfectly suited for journeys to the outer planets, whether by robots or humans, he was concerned that the Grand Tour could jeopardize his project in the process. Such are the hurdles that programs of space exploration must jump before coming to fruition, and it is a process that has changed little in the intervening years. Anderson was successful in cutting the Grand Tour's budget in May 1971, which resulted in Wernher von Braun, usually a staunch advocate of anything that would project a human presence into space, expressing his concern that the continued big-dollar funding of NERVA would endanger the Grand Tour in favor of a new rocket engine that, as he put it, had "no place to go."[3]

The Woods Hole group met again in 1971, with the assembled scientists voting twelve to one in favor of the Grand Tour, but characteristically adding that a cheaper alternative—in case the Grand Tour was not fully funded—that would send cheaper Mariner-based spacecraft to Jupiter and Saturn.

The debate raged on, and time was getting short—the planetary alignment that favored a tour of the outer planets would proceed whether Congress got its act together or not.

As NASA prepared its budget for 1973, yet another distraction entered the conversation—the funds required for the new space shuttle. This was the Nixon administration's desired replacement for the Apollo hardware, thought at the time to be cheaper and more reliable. While this would not be true in practice—the shuttle was expensive and, as it turned out, somewhat dangerous to fly—at the time it was the only game in town for human spaceflight. Once again, the Grand Tour felt the pinch. Within months, it was made official, when the then administrator of NASA, James Fletcher, announced that the shuttle would be moving ahead and that the Grand Tour would be replaced by the less expensive Mariner option.

Discouraged but undeterred, the engineers at JPL began to modify the existing Mariner spacecraft design for the outer planets. Quietly but doggedly, they designed systems that would not only accomplish their primary goals at Jupiter and Saturn, but that might also last long enough to fly past Uranus and Neptune if conditions—and ongoing operational funding—were favorable. They were not about to lose this opportunity to the bean counters in Washington, regardless of how few beans were left on the table.

As an official at NASA headquarters later said, "The lesson to be learned from Grand Tour cancellation was that you never fund such a big, long-term project at once. So we kept on adding piecemeal. And it's interesting that they always come out big. When you have less money, you can even do better sometimes."[4] JPL learned its lesson well, and in the next few decades would engineer spacecraft that would typically outlast their primary missions by years, and sometimes by decades.

John Casani recalled what occurred next:

> They canceled the mission, and Bud Schurmeier (who was the project manager before me), Ed Stone, and Dr. Pickering, went back to headquarters and said, "Hold on, we don't have to cancel the whole program. We can take the Mariner technology," which by now was a brand name, we'd had a number of successful missions, "and by just extending that technology a little bit, we can go to Jupiter and Saturn. We can't guarantee getting to Uranus and Neptune, but we can do a good mission at Jupiter and Saturn." So it was based on that premise that Mariner-Jupiter-Saturn was started in 1972.[5]

The Mariner design was evaluated in every detail. The engineers would design a new and reprogrammable onboard computer that would be capable of years of operation with a degree of autonomy. Batteries would be replaced by a long-lived nuclear fuel supply, and new science instruments would be designed that were relevant to outer planet research and capable of withstanding the intense radiation fields of giant Jupiter and Saturn—fields far more intense than anything encountered thus far.

By the time they were done, the Voyager program would be born, and the twin spacecraft would bear little resemblance to the Mariners they evolved from.

Chapter 21

PIONEERING JUPITER AND SATURN

As the work progressed on what would become the Voyager program at JPL, the science packages were being defined based on the forecast trajectory past Jupiter and Saturn. While these included instruments for the kinds of things you'd expect for surveying the outer planets—magnetometers, spectrometers, radar, and the like—it did not, early on, include cameras. While such a seeming oversight would be unthinkable today, at the time the debate still raged about whether cameras were worth the tradeoffs in weight and power consumption. TV cameras in the early 1970s were still large, heavy, delicate, and power hungry, and a lot of planetary science people resented the mass being devoted to something "trivial" like images from the gas giants.

As the program manager at the time, however, Schurmeier knew that images from distant and majestic planets had value both scientifically and beyond—JPL was discovering the value of public relations, a skill they would perfect over the next two decades. Regardless of the complexities, there would be cameras on the Voyager spacecraft.

At about the same time as this was going on, two quieter missions were embarking toward the same planets that the soon-to-be Voyagers would be exploring. Pioneer 10 and 11 were headed out to Jupiter and Saturn for the first-ever reconnaissance of those worlds, and would send back information critical to the design of the future Voyager probes.

The Pioneers were exactly that—newcomers to the neighborhood in every way, and traveling light. They were far smaller and less capable than the spacecraft that would follow them but were still capable of doing important science and pathfinding. The Pioneers were managed not by JPL but by the Ames Research Center, and were built by aerospace contractor TRW. The combined construction contract was a bargain at $380 million, when compared to the first estimates for the Grand Tour, and they even came with a warranty. As one wag of an engineer at TRW said, "This spacecraft is guaranteed for two years of interplanetary flight. If any component fails within that warranty period, just return the spacecraft to our shop and we will repair it free of charge."[1] Hardy-har-har.

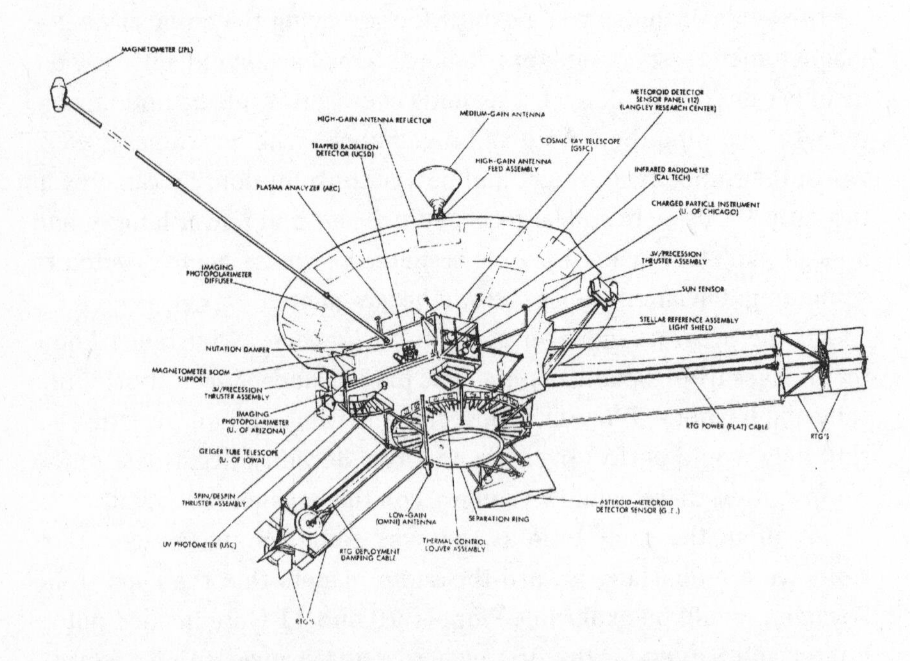

Fig. 21.1. A schematic of Pioneers 10 and 11. At the time, such programs still launched spacecraft in pairs to assure the best chance for success. It was often a good thing they did. Image from NASA/JPL-Caltech.

Each Pioneer weighed in at a meager 571 pounds, and each had four tiny nuclear power supplies that provided the craft about 155 watts of power (at the beginning of the mission; this declined over time)—about enough to light a high-intensity light bulb. The spacecraft were dominated by their nine-foot-wide radio dishes, with a smaller instrument cluster, called the bus, behind it. The computers were fairly basic, with most of the actual calculations performed on Earth and radioed up to the spacecraft to be stored and enacted by the computers as the mission functions came up.

Unlike the upcoming Voyagers, the Pioneers were spin-stabilized. This worked just like a spinning football in flight—the object continues to travel straight and true along the axis of the spin. The good news was that spin stabilization was a lot simpler than other kinds of deep-space navigation; the bad news was that it made it harder to perform certain operations, like taking pictures as one flew by the planets—the spacecraft were intended to be spun at a rate of five revolutions per minute, whirling like a slow top (Pioneer 11 actually spun faster due to a thruster malfunction after launch, at about eight rpm).

Pioneer 10 launched in March 1972 and Pioneer 11 in April 1973. A few months later, in December, Pioneer 10 encountered Jupiter, with Pioneer 11 making the same rendezvous a year later in December 1974. Pioneer 10 then headed on a path that would take it out of the solar system, while Pioneer 11 continued on to encounter Saturn in September 1979. By that time, the Voyagers were already on their way to the same locale.

Pioneer 10's closest approach to Jupiter was just over 82,000 miles—close enough to begin to understand the true intensity of the radiation in the Jovian system. At about 10,000 times what a spacecraft would encounter near Earth, the radiation was cooking the plucky Pioneer. In fact, the energy was so intense that high-energy particles started to generate false commands within the electronics of the small computer that had to be corrected via transmissions from Earth—these particles would pass through a computer chip and either burn out a tiny portion, or flip a binary one to a zero or vice versa. If you've ever experienced a corrupted file on your computer, you know what I'm talking about—flipped bits are bad juju.

With regard to imaging from the Pioneers, in a flash of engineering genius the mission used the onboard photopolarimeters—small devices used to measure the intensity of light, in this case in two colors, red and blue—to create images of the planets the spacecraft passed. Remember that these craft were spinning, so the photopolarimeter swept past the planet many times per minute. The data from these devices was collected, interpreted, and corrected for color, then distributed to the media in an approximation of actual color—JPL even created short movie clips of a rotating Jupiter from them. Hundreds of images were relayed by the twin spacecraft and, while they were low resolution and a bit fuzzy, caused a furor when released to the public. The images were also helpful to the promotion of the Voyager program, now underway.

Pioneer 10's last signal was received in 2003 as it was continuing toward the boundary between our solar system and interstellar space. Pioneer 11 flew past Jupiter in late 1974, skirting the planet at a comparatively tree-topping 26,600 miles. The rudimentary imager captured the planet's Great Red Spot, a cyclonic storm active for hundreds of years, as well as providing a first look at the polar regions of the planet. The huge planet's gravity swung Pioneer 11 onto a course for Saturn, which it reached in September 1979. Skimming over the cloud tops at an altitude of just 13,000 miles, the probe was directed to follow a path past the rings at a trajectory approximately similar to that planned for the Voyagers. There were concerns about how dense the ring particles might be, as well as how far from the rings they might extend, so Pioneer 11 blazed a trail for Voyager to safely follow. On the way out of the Saturnian system, the spacecraft imaged Titan, the planet's largest moon and an important target for missions to come.

Pioneer 11 was losing its ability to power the onboard instruments by late 1995. Regular contact was discontinued at the end of that September, with occasional peeks at the receding radio signal until 2002.

Of note, both the Pioneers carried plaques with representations of a human man and woman—the man has his hand raised in greeting (a gesture thought to be overly male-centric by some). It also contained a representation of a hydrogen atom in transition and a graphic indication of our sun's position relative to the center of our galaxy, with lines radiating to known pulsars, rotating stars that emit regular energy bursts. The final touch was a graph of the solar system, with a line indicating the route of the Pioneers past the planets and out of the solar system. Carl Sagan oversaw the creation of the plaques, and they were added late in mission planning. It was a last-minute choice to send a "message in a bottle" to any extraterrestrial life that might find the spacecraft adrift in space as to their points of origin.

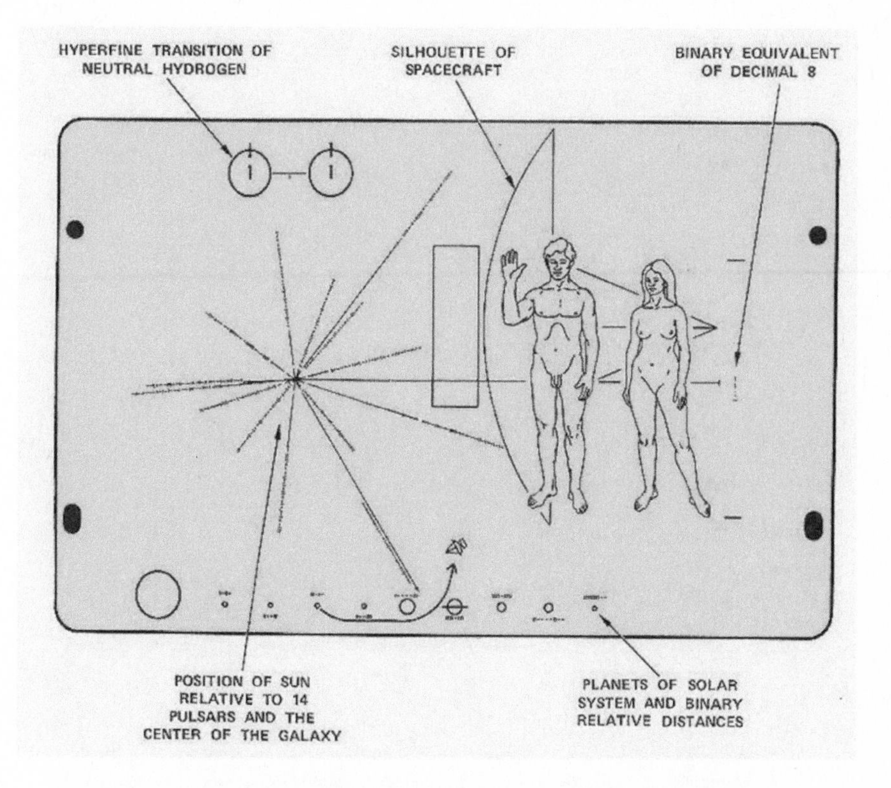

Fig. 21.2. The plaque attached to Pioneers 10 and 11 with explanatory callouts. Some taxpayers felt that the representations of a man and woman were "obscene," while others felt they were too Eurocentric, and that depicting the man with his hand raised was chauvinistic. Sometimes you just can't win. Image from NASA/JPL-Caltech.

Chapter 22

JPL'S FINEST HOUR: THE VOYAGERS, PART 1

As the early Jupiter data came back from the Pioneers in 1973, the results were interpreted and applied to the burgeoning Voyager probes, though at this time the program was still called Mariner-Jupiter-Saturn (MJS). The JPL crew was designing the probes for an extended mission, but that did not need to be made public—they would just build good, solid machines with the expectation of a long expected lifespan. JPL has been doing business in a similar fashion, though less clandestinely, ever since—witness the fifteen-year mission of Opportunity on Mars.

The news from Jupiter was harsh—the radiation was killing the little Pioneers. Based on this data, one Voyager experiment, an ultraviolet photopolarimeter, was deleted entirely and replaced with another, due to the radiation near the outer planets being over a thousand times what had been envisioned. As the instruments and experiments to be included on MJS were whittled down to their (mostly) final selections, NASA broke them into eleven areas of investigation and assigned teams to tend to the development of each. These teams were: imaging, radio science, infrared spectroscopy, ultraviolet spectroscopy, magnetometry, charged particles, cosmic rays, photopolarimetry, planetary radio astronomy, plasma, and particulate matter. This was one of the largest such efforts to date and would result in a well-refined spacecraft by the time the twin probes were launched later in the decade.

But the radiation problem went beyond simply parsing out the relevant experiments. The environment near Jupiter and Saturn was so hostile that a lot of work would have to go into designing systems and individual components that could survive the flyby. The MJS spacecraft were much more sophisticated and capable than the Pioneers had been, and with that came complexity. With the complexity came fragility—there would be many more things to go wrong.

John Casani had been moved into senior management of the mission, and he had strong concerns. He wanted to be sure that the spacecraft would be as tough and long-lived as possible. Who would know a lot about radiation shielding of delicate electronic components, he wondered? Aha! The military. "Most of the hardening work was ... funded by the DOD (Department of Defense). They were worried about the effect of a nuclear blast [on electronics] on the performance of avionics, aircraft, and what have you," he said.[1] The experience would benefit both the customer and the client.

A central focus of this shielding work was the computers, one of the more delicate systems of the spacecraft. Though it was still under development, there was not a lot of time to change things, and decisions would have to be made soon. "We had six computers on board," Casani said. These machines incorporated an early form of artificial intelligence, necessary since the spacecraft could be operating for periods of time without direct intervention from the ground. "The computers would cross check each other ... It was amazing. When the Grand Tour mission was canceled, we were in the midst of developing a computer called STAR, meaning self-testing and repair."[2] STAR had been one of the elements that caused the cancellation of the Grand Tour due to its projected cost. Fortunately, work that had already been done on the system could be integrated into the MJS mission. "When we came back and proposed to do something much simpler, just a mission to Jupiter and Saturn ... it didn't require a commitment of a large sum of money for new technology ... we all knew that the space craft would probably last well beyond Saturn."

But these amazing new thinking machines still needed protection. "As Pioneer 10 and 11 went past Jupiter," Casani said, "they found that the radiation environment was much larger than we had anticipated and designed for. We had to do some redesign and augmentation to account for that higher radiation, right down to actually putting little titanium hats on individual components. We also put a lot of effort into calculating what the effect of that shielding would be for every electronic component inside the spacecraft."

Fig. 22.1. John Casani's career at JPL has been long and storied. He is seen here at center, with Wernher von Braun to the left and James Van Allen to the right, looking over a model of Pioneer 4. Image from NASA.

The combined thinking power of the Voyager's computer was just over sixty-eight kilobytes. Some individual units, such as the one that monitored its health and drove basic flight functions, had only four kilobytes. Voyager's computer also had to be reprogrammable from the ground while in space for over the course of a decade or more—much more demanding than the Apollo computer, which had a flight life of under two weeks (at the most) and in which commands were entered via an astronaut's fingers on a keypad. So rather than using the hard-wired memory systems like Apollo and the early Shuttle computers, Voyager used rewritable memory, which came in the form of a sophisticated version of an 8-track tape recorder of the day—a very expensive version of the one in your father's 1968 Mustang. That gave it about one-hundred-thousandth the computing power of an eight gig iPod, for those of you who even remember iPods. In today's world it would barely qualify as a reasonably capable calculator ... and yet, beyond our solar system, this underpowered machine continues to function well into the twenty-first century.

One of the computer's key tasks would, of course, be guidance and navigation, and to do this successfully, it would have to know where it was at any given moment. The location could be tracked using gyroscopes and information from Earth, but the spacecraft would still need to be capable of taking navigational readings from the sun and other stars to know where it was. This was much more complicated in the outer solar system than it had been near Mars or Venus, because, as you might expect, it's *dark* out there. "Someone came up to me one day with a technical briefing saying that they were worried about the sun sensor because it would barely have enough sensitivity to get us to Saturn. I said 'What are you talking about?' Well, the light from the sun would be so dim out there, and at that time, output signals from the detectors was pretty noisy."[3] Casani was referring to electronic noise in the system, which might be sufficient to overwhelm the information the probes needed to navigate properly. Casani continued: "I said to him, 'That's not acceptable, you need to come up with a fix that will not only get us out to Saturn, but will take us out past Neptune as well.'" Fortunately, they did.

Previous sun sensors had been pretty simple designs—essentially two photocells with a little baffle between them. If more light shone onto one cell than the other, the spacecraft would steer to compensate, back and forth, until the measured light was about equal. But with the sun as weak as it would be out by Saturn and beyond—just an extra, extra bright star at that distance—it would be dodgy. "You can only extend that kind of design so far, and it wouldn't quite make it to Saturn," Casani said—the sun was so much dimmer at that distance. "So we had to put some electronics and amplifiers and things like that into the system. Once you do that, you're into a more complicated device. . . . The problem with doing that at the time was the cost."[4] But in the end they figured it out, and by hook or crook, they would make it to the edge of the solar system.

But the computer still had to be capable of some solo work, and this was an ongoing problem.

"We just kept peeling that onion back and addressing everything we could think of," Casani said. "There are such long communication times, and if something happens in the spacecraft . . . you have to have onboard system either smart enough to be able to figure out what went wrong and how to compensate for it, or you have to tell the space craft to go on in quiescent mode—just perform at a bare minimum to operate. We call this safe mode, in which the computer waits for the ground to try to figure out what to do."[5]

There would be periods where the spacecraft would not be in communication with the Earth, when it was handing off from tracking station to another, or when it was passing behind a planet or a moon, for example. Unless something was blocking access to the sun (the same general direction as Earth), the spacecraft's radio antenna was always supposed to be pointing toward home. "If it was not pointed at the sun, it knew that," Casani said. "It would try an automatic recovery routine and sun search, and the sun acquisition routine had automatic triggers if was not." With the sun in sight, it still needed a second star fix to know its orientation. "The other degree of freedom is the rotation about this sun axis, and that's what we use star trackers for. It would rotate until a recognizable star pattern came in the field of view—in this case, Canopus—and then fix on that." This was still rocket-ace stuff in the 1970s.

They had swatted down most of the major issues that could cripple a mission, but concerns about the radiation-induced problems soon came back to the fore. All that radiation might have another side effect, the engineers soon realized—static electricity. Casani continued:

> Things were going along pretty well, but then the people who were doing the analysis of the Pioneer data were coming up with new understandings and insights, and one of these was that while going through the radiation belt there was the possibility of inducing a static electricity charge in the cables and external structures of the spacecraft. In theory, thousands of volts might be conducted into the spacecraft! So we went through it in detail, and in retrospect what was a bit of a comical situation. Not having time to do anything fancy, we did the only thing that seemed practical, and that was to go out and buy a bunch of aluminum foil and wrap the cables and some of the electronics in that. We used rolls and rolls. One of the engineers pointed out later that what had been flying on our spacecraft and protecting it from radiation is the same foil you've got in your oven for your Christmas turkey.[6]

Gobble-gobble, Voyager.

With the engineering in hand, someone needed to coordinate and oversee the science side of the program. Ed Stone, a Caltech physicist who had been with the Grand Tour project since 1970, was appointed the project scientist—the big kahuna for Voyager. Notably, he's been working on the project ever since, continuing to go over the mission's collected data to this day. In this position he was responsible for overseeing all the team scientists and mediating between them—an important but unenviable task—and between the overall science effort and NASA's budgetary constraints—an even less enviable task. If the budgetary numbers—whether they be related to mass, power, or dollars—demanded the alteration or removal of an experiment, he would be the bearer of the news to that team leader and would hear the inevitable anguished response. As Stone put it, in terms only a scientist could love, the project scientist served as "an impedance matching function between the engineering requirements and constraints and the science requirements and constraints to try to find a way to achieve the optimum match between these two different sets of requirements and desirements."[7] (And yes, desirements is a real word.) Stone became the matchmaker of the Voyager program, blending requirements and desirements as well as could be humanly accomplished, while continually keeping his eyes on the prize, a maximized science return.

Fig. 22.2. Ed Stone teaching at Caltech in 1972, the same year he was appointed as the project scientist for the Voyager program. Image from NASA/JPL-Caltech.

Stone soon realized that the job was more involved than he'd thought when he accepted it. One of his duties would be to choose between different sets of similar observations and let the scientists suggesting them know which one would be made and which one would not, based on mission constraints—this was a flyby mission, not an orbiter, so time would be limited at each target. "It turns out, that's a much more critical role than I had thought ahead of time, and that's because ultimately what science is all about is making discoveries. By deciding to make this observation rather than that one, you're effectively deciding that that group of scientists gets to make a discovery and this group doesn't," he said.[8] At times, it surely felt like having a bullseye painted across your back, head to toe.

And then, there was the public-facing role. While discoveries were still written up for publication in peer-reviewed journals as they always had been (and still are), the announcements of those discoveries to the media were a more immediate affair, and with that immediacy came human emotions. Scientists usually like to hold their newfound data close to the vest while working through it, and then report their findings, with the attendant rights of ownership (and the career-building recognition that can come from it) when complete. But with a publicly funded, fast-track mission like this one, there was no time for such a process. Stone felt that not sharing data immediately with the rest of the team, and ultimately with the world, "would have inhibited the total development of the scientific program."

This impression came to Stone during a press conference for Pioneer 10:

> Here was a room full of reporters wanting to know what the scientists had discovered. I mean, to me that was incredible. Normally there just isn't that much interest in what you're doing as a scientist. And here they were day after day saying, "Tell us what you've discovered. Tell us what you've discovered." I realized that with Voyager we had both the opportunity and the obligation to communicate what we were discovering; to help the media tell the story. But we had to do it in a scientifically credible way.[9]

He realized that the public was galvanized by the results coming in from the frigid reaches of the solar system, and besides the scientific value of sharing there was a huge advantage to be gained by sharing discoveries with the world as soon as reasonably possible. He decided then and there that the people on the science team would be just that—a team—and they would share as a team as well.

"We did go through a process of some real-time peer review during the encounters," Stone noted,

> And this allowed us to bring the public along on the journey. Day by day, they could actually see the scientific process of discovering things you hadn't expected or seeing things that you could not understand, at least initially. The encounters really brought the world media to

> JPL. You had to be there, the only way to get the pictures in those days was to be there and get a hard copy. This was the only way to get a digital image across the country—we could get digital images back from the planets but we couldn't get them across the country! So over a ten-year period, starting with Jupiter in 1979 and Neptune in 1989, the media group became a part of the team. Often, rather than having questions, they would have suggestions . . . why not try this or why not do that? They were learning just how different and diverse everything was, right as we were seeing it.[10]

As the spacecraft design matured, and with them the mission design, it was clear that while this may not have been the Grand Tour as originally envisioned, neither was it just another pair of Mariner flights. This was big; a once-in-a-lifetime chance to see most of the solar system in one sweep if all went well. In early 1977, before the first launch, the name was officially changed to Voyager.

Voyager 2 launched in August 1977, and Voyager 1 in September. The problems started right away, during the ascent of Voyager 2. The new computers had a sophisticated fault detection and correction system, and were activated just prior to launch. But the engineers had forgotten one thing—to tell the computer just how wild a ride launch could be. "This was a bit of a screw up," Casani said. "During ascent, the launch vehicle needed to roll onto its back to set up the proper trajectory. That roll rate was larger than we had designed the attitude control gyros and inertial control equipment to handle. It would never have seen anything that large in flight. The net effect of that was that it saturated the gyros."[11] In effect, it drove the internal gyroscopes, a critical part of the navigation system, past their limits. "We had three gyros, and at any given time we only needed to have two gyros operating. If any one of them failed, the third would be directed by the computer to take its place. But when the roll rate exceeded the limitations, the gyros saturated. When this happened, we lost the signal from both axes. We hadn't allowed for that." The computer was, to re-appropriate a phrase, *lost in space.*

Casani continued: "So the system saw that we had a failed signal from one gyro and decided to switch to another. But then that one had the same problem; it then started switching computers, then processors, then power sources—there were six levels of change-outs of hardware that the system went through. On the ground, we could not figure out what the hell was going on. We assumed that the system was responding to a real failure."[12] But it wasn't a failure, it was just the gyros being overdriven. "I think it was a day and a half before everything settled out, and the guys on the ground were trying to figure out what to do. But the guy who was the architect of the fault protection system, Chris Jones, said 'I think this is doing exactly what it's supposed to be doing . . . let's just hold on.' And sure enough, once everything settled down, the system restarted itself and everything was fine. As it turned out, all we had to do was to do *nothing* and the spacecraft took care of itself."

Then Voyager 2 had a radio problem. The computers were programmed to check for a signal from Earth every twenty-four hours, and if that signal was not detected it would initiate a series of procedures to fix what it assumed could be a problem. Of course, you have to be transmitting from Earth for the radio to receive a signal. "We were chugging along," Casani said, "and due to an operational error on the ground, a timer was not set for the next interval." Someone on the ground had forgotten to tell the transmitter to send a signal up to the spacecraft to let it know that everything was fine. When Voyager 2 did not hear the signal from Earth, it assumed that there could be a problem with the onboard radio receiver and turned on the backup radio. "But as soon as it did that, just like sometimes incandescent light bulbs in your house fail right when you switch them on, the other radio failed right away. That was an attention getter. So we managed to switch back to the first radio, and then *it* wasn't working properly. The problem was that it had a tracking receiver that checks the ground frequency. That's necessary because the ground frequency is constantly changing due to Doppler effects from the rotation of the Earth and the velocity of the spacecraft." The frequency of the signal coming up from Earth actually shifts due to both the rotation of our planet and the speed of Voyager as it moves away from Earth. But when it did not see what it expected (due to an electronic fault), the computer decided that, once again, there must be an onboard problem, and shifted frequencies to compensate. "The receiver just reset itself to some nominal setting on its dial," Casani said, and it was now unable to pick up radio transmissions from Earth. "So we had to slide the ground frequency until we found where the spacecraft was tuned, then keep track of that." But this frequency would continue to shift as the spacecraft proceeded in its journey—the proper frequency was a constantly moving target. "So the team came up with a way of calculating what the Doppler shift would be, and programmed that into [the transmitters on Earth] to automatically shift the ground transmitter to account for that shift. That happened on the way to Jupiter, and that receiver has been working in that mode for forty years."

There were other changes engineered into the system as the Voyagers proceeded through their missions as well, which demonstrated the incredible utility of having a robust and reprogrammable computer onboard. One of these had to do with imaging dim planets, Stone said: "When we launched, we did not have the capability of returning any images from Neptune—it was just too faint. That was all accomplished after launch. We reprogrammed the computers to take the smear out of the images by maneuvering the spacecraft the right way, so that in the very long exposures of Neptune, the images didn't blur."[13] They would maneuver the Voyager as it was taking the image to track the planet as it whizzed by, effectively keeping the camera in one place as it took the picture.

Although launched after Voyager 2, Voyager 1 took a faster route to Jupiter and arrived there in January 1979. During its fast flyby, the most intense period of the encounter was the two-day period approaching and departing its closest approach to the giant planet, as it evaluated the environment and imaged Jupiter and some of its moons. Jupiter's magnetic field and radiation were measured, and a faint ring was discovered surrounding the planet. This last item was a surprise; it turned out to be composed of dust from nearby moons, due either to collisions or to some unknown event in the distant past. Voyager 1 first spotted the rings, then Voyager 2 was programmed to do a more thorough investigation of the ring system—yet another advantage of twin probes in a mission such as this.

But the big surprise was Jupiter's moon Io—sometimes mistakenly called "Ten" by inattentive members of the press who had trouble discerning the small letter *o* from a zero. Voyager 1 passed Io at a distance of just over 12,000 miles, resulting in some stunning images and some remarkable discoveries. As Stone put it, "Before Voyager, the only known active volcanoes were on Earth. Then we flew past Io, which is about the size of our moon, which had ten times the volcanic activity of Earth!"[14]

The story of the discovery of these volcanoes is one of serendipity and perseverance. A navigation engineer on the Voyager project, Linda Morabito, was looking closely at images from the Io flyby. She spotted a weird flare off to one side and eventually realized that it was a volcanic plume reaching about 170 miles into Io's sky. "By February 1979, the data was falling down on us like rainfall and the images were coming in at all hours of the day and night," she said.[15] "On the morning of March 9, I arrived at my station and began processing several images taken by the Voyager 1 spacecraft as it was looking back over its shoulder for one last view of the Jovian system." The images were sensational, as were all the pictures returned for the mission, but something didn't look right to her eye: "I suddenly noticed an anomaly to the left of Io, just off the rim of that world. It was extremely large with respect to the overall size of Io and crescent-shaped. . . . Ed Stone, the Project Scientist, came down to look at the image and I remember the absolute wonder on his face. . . . [He]very quietly said 'This has been an incredible mission,' and he repeated that several times. When Ed left, the work began to correlate the anomaly with the surface of Io." A few days later, while having dinner with her parents, her father suggested to her that she may have discovered the first volcanic activity beyond Earth—and, of course, that is exactly what she had achieved.

Researchers think there may be as many as four hundred volcanoes on Io, the result of being squeezed by the titanic forces of Jupiter's gravitational field and those of the other large moons, Europa, Ganymede, and Callisto. This squeezing and scrunching causes Io's interior to remain hot due to internal friction, resulting in extreme volcanic activity. Plumes of sulfurous material have been spotted shooting as high as three hundred miles above its surface. Io is also covered with mountains, some higher than Everest on Earth. Unlike many outer planet moons, which have icy crusts, Io is rocky, with a silicate crust and iron or iron-sulfide core, just like the terrestrial planets.

Io was indeed a serious surprise, not just due to volcanic activity but also because of its appearance. The place looked like a pizza that had been left outside for a few weeks—a malignant orange and yellow, with dark patches of gray and dark brown. "It looked like a moon with some kind of skin disease," one researcher said.[16]

But Io is more than just an ugly moon—it is also a prime contributor to the pollution of the Jovian magnetosphere. Sulfur, oxygen, and sodium are constantly ejected from Io and swept up into the intense fields that surround Jupiter. Io is also rich in water, and it is the densest of Jupiter's moons. It is so geologically active that there were few craters observed on its surface—it simply resurfaces itself too often.

As the spacecraft departed the Jovian system, it continued to image Jupiter. There were patterns emerging from the complex, swirling clouds that spread across the planet's surface. What looked like sharp bands from Earth-bound telescopes, and even on the Pioneer images, were now shown to have complex, swirling interactions at their boundaries.

The ability to photograph these over a period of time, even the comparatively short period of a flyby, indicated convection occurring at these boundaries—upwelling and downwelling of gases from the interior. The Voyagers also observed the Great Red Spot, a cyclonic storm first seen in 1830 and possibly in earlier telescopic observations. The storm's wind speeds were determined to be about four hundred miles per hour, across an area twice the width of Earth, and it moved counterclockwise, appearing to make a complete revolution in four to six days.

The Voyagers also spotted auroral emissions in Jupiter's polar regions, another strong indicator of its magnetic field, as well as super-bolts of lightning across its cloud tops. Peering into that atmosphere, the probes measured temperatures ranging from -170 degrees Fahrenheit to a broiling 1,500 degrees Fahrenheit deeper inside the atmosphere.

Both spacecraft also imaged Jupiter's moon Europa, just slightly smaller than Earth's moon. Voyager 1 grabbed a series of long-range images showing intersecting linear features that looked like cracks in the surface, but as Voyager 1 left the Jovian system Voyager 2 made a closer pass by Europa, and the next set of images showed little vertical relief near these features—cracks should exhibit some deformation from their forceful creation. One puzzled mission scientist commented that it looked as if they "might have been painted on with a felt marker."[17] Europa also experiences tidal pressures and resultant heating similar to Io, but at only about one-tenth of the intensity due to its being further away from its giant, bullying parent planet. Europa also sported few craters, suggesting a highly active geological history.

And then, off the twins sped toward Saturn, over a year away. The quick transit past Jupiter and its moons was a scientific smorgasbord, just a sampling of what they could accomplish with orbiting spacecraft. "When you're in orbit, you have the chance to redo things," Stone later commented. "If you learn something, you can say 'ok, let's go and do this in the next orbit.' In a flyby, you have to plan everything ahead, you have just one chance, Of course, in the case of Jupiter and Saturn we did have a second chance after Voyager 1, because Voyager 2 came along a bit later, and we did change some things on Voyager 2 as a result of Voyager 1. When you're flying by, you have to make a decision about where you *think* the discoveries are going to be, and hope that you pitched the right way, because the spacecraft is going to fly right by and onto the next planet. The other thing that's different from an orbiter is that you would have these periods of getting ready, planning for the next encounter, planning second by second what we wanted to do while still analyzing data from the prior encounter. It was a very busy time," he concluded with a wry smile.[18] It was indeed busy, and intensely rewarding. Many JPLers still recall their time on Voyager as one of the most exciting parts of their lives due to the intensely busy schedule and the forced efficiency of flyby observations.

Fig. 22.3. From left to right, the author, Ed Stone, and John Casani in 2018. Image from AIAA.

Chapter 23

BRIEF ENCOUNTERS: THE VOYAGERS, PART 2

Flying past Saturn was a true showstopper—as the most visually spectacular planet in the solar system, Saturn promised not just great science but incredible views. Its beauty aside, the science just kept coming. "Saturn, like all of the planets the Voyagers visited, was full of exciting discoveries and surprises," Ed Stone said.[1] "By giving us unprecedented views of the Saturn system, Voyager gave us plenty of reasons to go back for a closer look." It was a perfect precursor to a follow-on orbiter mission.

Voyager 1 catapulted through the system in November 1980, passing within 77,000 miles of the planet, while Voyager 2 followed in late August 1981, at a close pass of about 81,000 miles. The imagery coming back from both spacecraft defied description; the Saturnian system is breathtakingly beautiful in all respects.

The Voyagers made good use of their brief time at the planet. Wind speeds at Saturn's equator were gauged at up to 1,100 miles per hour. The planet had more hydrogen in its upper atmosphere than Jupiter did, suggesting that the dynamics of the two gas giants were somewhat different—in fact, Saturn is almost entirely made up of hydrogen and helium, and is the lightest of the planets, being lighter than water. Auroras were visible at the poles, as they were at Jupiter. Saturn was colder than Jupiter, with cloud-top temperatures a chilly -333 degrees Fahrenheit. Surprisingly, after what they had observed at Jupiter, the temperatures deeper inside Saturn's atmosphere were still cold at

-154 degrees Fahrenheit, a far cry from the heating observed below Jupiter's clouds.

Fig. 23.1. Saturn as imaged by Voyager 1. Better views have come from Cassini, but few were more dramatic. Image from NASA/JPL-Caltech.

But the real star of the show at Saturn was, of course, the rings. First observed by Galileo in 1610, they had puzzled scientists since that time. Some good data was returned by the Pioneer probes, and while these results raised as many questions as they answered they helped to identify observation targets for the Voyagers.

It was long presumed that the rings were composed of ice and dust particles, and were probably the remnant of unformed mass orbiting the planet—in other circumstances, it might have formed a large moon instead. The complete story of the formation of the rings, and in particular the gaps between them, remained unresolved. Some features were observable from Earth, while more became obvious as the Voyagers returned images back to JPL. The so-called Cassini Gap, the largest in the ring system, had long been suspected of being "cleared" by a moon that might be pushing aside all the ring matter in its path, but no moons large enough to have swept the wide ring-gap clear were found.

Fig. 23.2. Saturn's rings as imaged by Voyager 1. It's hard to believe that they are mostly ice crystals and a bit of sand. Image from NASA/JPL-Caltech.

More gaps in the rings were spotted, and they appeared to have variations in their width. The gravitational forces of the planet's many moons appeared to be the cause. A thready, irregular ring, called the F ring (they were designated alphabetically in order of discovery) was identified, and appeared to be comprised of three separate thin rings that were twisted and braided into one. These were thought to be caused by "shepherding" moons within the rings, their gravitational effects disturbing what would otherwise be regular, circular, and concentric rings. The images were truly jaw-dropping.

Dark "spokes" were also seen in imagery of the rings and seemed to correlate with Saturn's magnetic field. These spokes changed color as the angle of light radiating from them changed when the Voyagers flew by, starting out as darker than the surrounding rings and eventually becoming brighter than surrounding material as the spacecraft departed the system. More information would have to wait until an orbiter could be sent to loiter in Saturn's system—the much later Cassini mission.

Of Saturn's many moons—it is now known to have at least sixty-three of them—Titan was of primary interest. It was so interesting that Voyager 1 was specifically redirected to do the only close pass, a rerouting that would cause Titan to be its final encounter in the solar system—the gravitational effect as Voyager 1 sped past the moon at a closest distance of just 4,000 miles slung it northward relative to the solar system, sending it on a fast track to interstellar space.

"Voyager 1 completed the Jupiter-Saturn leg of the mission and [reconnoitered] Titan," Stone said. These were the original mission objectives. "That left Voyager 2 in the plane of the planets so that it could head on to Uranus. That was called the Voyager Uranus-Interstellar Mission, because Voyager 1 was now headed into interstellar space. There would be no further planets for Voyager 1."[2] Voyager 1 "took one for the team" by flying past Titan, allowing Voyager 2 to continue its journey to the outer planets.

Besides being the largest moon of the planet, Titan was also the only one known to have a palpable atmosphere—dense, orange, and impenetrable by visual means. Larger than the planet Mercury, 3,200 miles in diameter, Titan is far larger than Earth's moon by over a thousand miles. Until the arrival of the Voyagers, it was thought to be the largest moon in the solar system, but it lost this title to Jupiter's Ganymede by a few hundred miles once the smaller size of the rocky body beneath the atmosphere was understood.

And then, there was the atmosphere itself to be understood. As had been done at Jupiter and Saturn, Earth-bound observers carefully monitored Voyager 1's radio signal as it passed behind Titan, to measure the density, pressure, and basic composition of its gaseous envelope. "Before Voyager, the only known atmosphere with nitrogen was here on Earth. Then we flew past Titan, and it has a denser nitrogen atmosphere than here on Earth. Our view of planets and moons in the solar system was really stretched by then," said Stone. The composition of the air surrounding the planet was not entirely unlike the dense smog that enfolded the Los Angeles basin during the Voyager program, filled with hydrocarbons. This led to speculation that the surface might have vast lakes of liquid hydrocarbons—ethane and methane—a hypothesis that would have to wait for the Cassini probe in the following decade. The average temperature was hardly equivalent to a hazy Los Angeles afternoon, however, at a frigid -290 degrees Fahrenheit. The further observation of how Voyager's trajectory was influenced by the moon gave scientists a good measure of Titan's mass as well.

The moon Enceladus was also intriguing, appearing to have a complex and young surface, partly cratered and partly smooth. The smoother regions were estimated to be far younger than the cratered ones, less than a few hundred million years for some of them. Faults crossed other regions. Tidal heating was suspected in the formation of these cracks in the surface, but, again, further investigation would have to wait for a later mission to the Saturn system.

There are many tales told by the other moons of Saturn, but one more deserves mention here: Mimas. This is not due to its size; it's only about 246 miles across. And it's made mostly of water ice with a bit of rock, so it's no record holder there. But it is emblazoned with a massive crater called Herschel, named after the famous astronomer who first spotted the moon as a pinprick of light in 1789. Herschel (the crater, not the man) is 81 miles in diameter, almost one-third of Mimas's diameter, and that's a big hole for a moon about the same size as Spain. At its deepest, the crater is six miles below the surrounding surface, with a marked central peak that rises four miles above the crater floor. The impact that made the crater was so devastating that there are fractures on the opposite side of the tiny moon. This is just one more example of why Earth needs a robust space program to allow us to divert asteroids from hitting our own planet (you know the dinosaurs wished they'd had one when that enormous asteroid slammed into Central America and put them out of business 66 million years ago).

Fig. 23.3. Mimas as seen from Voyager 1. The enormous crater, Herschel, is named after the moon's discoverer. Herschel is 81 miles across—a major bit of damage, given that Mimas has about the same land area as Spain and is about 246 miles in diameter. Image from NASA/JPL-Caltech.

As Voyager 1 blasted out of the solar system at almost 40,000 miles per hour, thanks to that final swing past Titan, Voyager 2 began its long, lonely trek out to Uranus, 1.7 billion miles from the ever-dimming sun. It would take five years to reach that aquamarine world, but it was well worth the wait, for when Voyager 2 passed the cold gas giant at a distance of 50,500 miles, there were many more surprises in store. Eleven new moons were discovered orbiting Uranus, making for a total of fifteen. The known moons, including Triton and Miranda, were examined at a moderately high resolution; the smaller bodies were imaged as well as was possible. Miranda turned out to be as interesting as the planet it circled, though in a much different fashion. Miranda is a nightmare of geological processes, with canyons that plummeted to a depth of twelve miles, and a patchwork of both young and old rocky surfaces. It appears, though this has not been confirmed, that at some time in its long, cold life, Miranda smashed into another world and was to a large extent shattered, then slowly re-aggregated (or coalesced) back into the moon we see today.

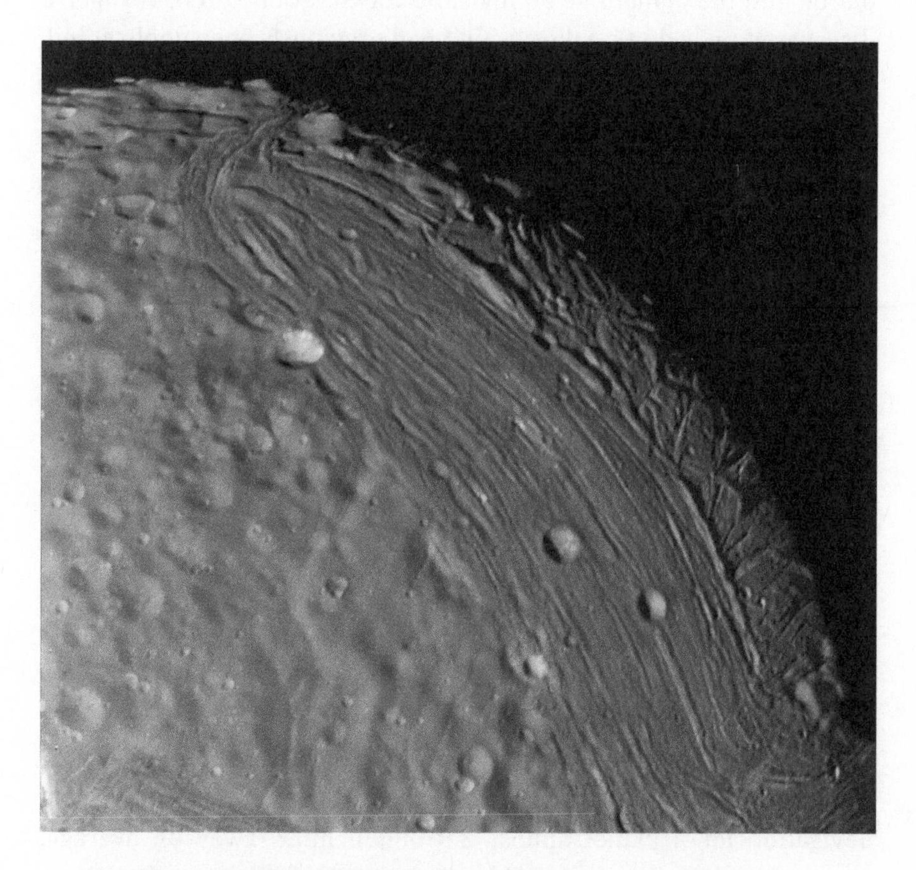

Fig. 23.4. Muddled Miranda, moon of Uranus, imaged by Voyager 2. Miranda's violent history can be inferred from the striking contrast between the cratered landscape to the left, compared to the smooth striations to the right. Image from NASA/JPL-Caltech.

The third-largest planet in the solar system, Uranus is an oddball in one sense—it rotates at a nearly 90-degree tilt from other planets, with its axis almost in line with the ecliptic, or nearly the same plane as the path of the sun in Uranus's sky. This strange tilt is thought to be the result of a collision with another large body early in the history of the planet—the outer solar system was a demolition derby in its early days. This tilt also results in a bizarre magnetic field that spews out behind the planet like an invisible corkscrew. In 1986, Voyager 2 discovered two additional rings that had previously been unobserved, bringing the total to eleven. Today we know that Uranus has thirteen known observable rings, along with a radiation environment roughly equivalent to that of Saturn's. The outer planets were turning out to be even less hospitable than had been thought—and they were already viewed as a pretty poor neighborhood for humans, albeit with a spectacular view.

The temperatures of the gas planets continued to drop, with Uranus showing an average of about -350 degrees Fahrenheit. Voyager 2 continued on.

The final act of Voyager's planetary tour occurred in August 1989. The spacecraft was steered out of the plane of the ecliptic and into a trajectory that took it over the north pole of Neptune. This trajectory change was required to allow the probe to curve past the planet and descend to the orbit of Triton, the largest of Neptune's fourteen moons. This final course adjustment would also determine the angle at which Voyager 2 departed the solar system into interstellar space, as Voyager 1 was already en route to do.

For this last planetary encounter, JPL controllers sent Voyager 2 skimming the top of the planet at less than 3,000 miles—pretty daring navigation for a planet almost 2.8 billion miles away on average. Among its other discoveries, Voyager imaged what became known as the Great Dark Spot on Neptune, similar to Jupiter's Great Red Spot. For an idea of scale, Earth would fit comfortably inside this dark patch. Rather than being a hundreds of years old cyclonic storm like the Red Spot, however, the Great Dark Spot appears to have disappeared over the course of about twenty years—more recent Hubble Space Telescope images don't show it.

To acquire the images of the furthest planet from the sun (Pluto was demoted to "dwarf planet" status in the twenty-first century), the flight controllers had to once again steer Voyager 2 as it passed Neptune to allow the cameras to track the planet as the spacecraft flew by to avoid image smear. It worked magnificently, with stunning images of the azure-blue world galvanizing the world press. Pretty impressive work, considering that Neptune only receives about three percent as much sunlight as Jupiter does.

Like Jupiter, Saturn, and Uranus, Neptune is another gas giant, with an atmosphere composed primarily of hydrogen and helium—think of Neptune and Uranus as "unfinished planets," the fixer-uppers of the solar system, maintaining more of the primordial gases they were formed from. While hauntingly beautiful, any solid surface is thousands of miles deep, under crushing bands of intensely cold gaseous atmosphere. Like Uranus, Neptune had plenty of methane in its atmosphere as well, which preferentially absorbs the longer wavelengths of light (on the red end of the spectrum), leaving it to reflect bluish light.

The upper layers of the atmosphere were measured at about -360 degrees Fahrenheit, but, as opposed to chilly Uranus, Neptune warms substantially as one moves deeper into the atmosphere. This heat drives the huge, cirrus-type clouds that encircle the planet. These are formed above the layers of methane-dominated atmosphere below and appear white in photographs. Judging from shadows cast by these clouds, the science team estimated them to be about twenty to thirty miles above the layers below. At the equator of Neptune wind speeds were estimated to be about 1,500 miles per hour, the fastest in the solar system.

Neptune has a strong magnetic field, but its source appears to be offset from the center of the planet by about 8,500 miles. This magnetic field varies as the planet rotates due to this offset—the pressure of the solar radiation from the sun pushes it back into a comet-like tail that changes shape over time. Even at its high rate of speed, it took Voyager 2 a full thirty-eight hours to exit Neptune's magnetosphere.

Just five hours after passing the Neptunian pole, Voyager 2 whizzed past Triton at a distance of about 25,000 miles. This last hurrah of its tour of the solar system did not disappoint, as Triton turned out to be another oddball moon. About three-quarters the size of Earth's moon, or 1,680 miles in diameter, Triton circles Neptune in a tilted orbit, and does so backward—the opposite direction of Earth's moon and most others in the solar system. It appears to be another geologically active world, with icy geysers spewing nitrogen gas and dust high above its surface, with a frozen nitrogen surface. Given its weird orbit, there is speculation that Triton may have been a captured moon, rather than having originally formed next to its parent planet. It has a thin atmosphere that extends about 500 miles above its surface, with a surface temperature of almost -400 degrees Fahrenheit. Triton was the coldest place yet measured in the solar system, as the exploration of Pluto was still almost twenty-five years away (Triton and Pluto turned out to be roughly equivalent in temperature).

Before it left Neptune's neighborhood, Voyager snapped images that would net six more moons circling the planet, all of them as dark as charcoal.

Finally, three dim, diffuse rings were found to be encircling the planet. The particles that they are composed of range from the size of smoke to larger icy bits. They are not visible from Earth, even to the largest telescopes.

And that was the end of the road, so far as the planets were concerned. After twelve years of amazing science, brilliant engineering, and exciting discoveries, Voyager 2 screamed away from Neptune and toward the boundary between the solar system and the great beyond. It is headed to an entirely different portion of the heliosphere—the vast bubble of solar radiation that is emitted from the sun—than Voyager 1 transited. Both spacecraft were now in the final portion of their journeys, the Voyager Interstellar Mission.

CHAPTER 24

INTO THE VOID: THE VOYAGERS, PART 3

If you go wandering around JPL looking for Voyager mission control, you won't find it. The control center for that daring program was evicted from the main campus of the NASA field center many years ago—office space was needed for new, higher-priority missions. So you will need to hop back into your car, drive down the hill and into Pasadena, and, as you enter the city, right at its border with neighboring Altadena, you will see some low, beige, nondescript office complexes. Entering one of those buildings, which could easily be (and probably, at one time, was) an office for a medium-sized life insurance office, you will find yourself in Voyager mission control.

A handful of people work in the building, in a few offices and some familiar felt-lined cubicles. It's pretty clear that the office décor is JPL surplus. In the center of the room, operating on an assured power supply, are a few old Sun workstations on some banquet tables—this is the Earth-node of Voyager. To these elderly computers flow the incoming data from the Voyagers as they continue their journeys, now over forty years in the making. It's not a very impressive sight—there are no big racks of blinking lights or pretty overhead screens displaying the graphics like they have over at the SFOF—but even though the visuals are not stunning, what's zipping around inside those computers is. Because inside those obsolete computers lives the Earth-side brains of humanity's first foray beyond the solar system, and it doesn't get much cooler than that.

Fig. 24.1. Until recently, Suzanne Dodd was in charge of the Voyager Interstellar Mission, and she is seen here at her desk in the offsite "mission control" for Voyager. She's been awarded more NASA medals and commendations than you could carry with both hands. Image from NASA/JPL-Caltech.

Although initially trailing behind Voyager 2 and the Pioneer probes, Voyager 1 was propelled so quickly through space that it arrived at the edge of the solar system first. It passed Pioneer 10 in 1998, and encountered a region called the "termination shock" in 2004, the area where the solar wind—charged particles from the sun—begin to slow down as they come up against forces from interstellar space—the region "outside" the bubble of solar radiation from the sun.

Suzanne Dodd, who was in charge of the Voyager Interstellar Mission at the time, explained: "The termination shock occurred 2004 for Voyager 1 and 2007 for Voyager 2. The spacecraft was now in a region called the heliosheath, the huge boundary layer where the sun's influence waned and that of the stars begins."[1] The region doesn't begin suddenly; this is a huge transition zone. The solar particles moved slower here than they did just a few months earlier on Voyager 1's path—about 250,000 miles per hour as opposed to one million mph observed just a few months before. Then, seven years later, the scientists noted another shift.

"The mission had been quiet until about 2011, with not a lot going on, until the low energy charged particle instrument started seeing changes in the direction of the sun's charged particles," Dodd explained.[2] "We'd thought these would bend and go back around, but what we saw instead was that they just stopped, and did not really go in any direction. This was named the stagnation layer"—kind of like the doldrums that sailors encounter in the equatorial regions of the Earth's oceans, where the wind can suddenly stop blowing for weeks at a time. "Then in mid-2012," Dodd continued, "we saw a drop of 50 percent or more of the number of charged particles from the sun. This went up and down about five times in the space of a month, and then by the end of August, just stayed down and has never come back up."

It turns out that this region that interfaces between the solar wind and what lay beyond was *rumpled*, and Voyager moved in and out of areas still abuzz with solar particles until finally exiting them altogether. The spacecraft was now in interstellar space.

"You have a spacecraft going where none has ever been before, and you expect to see something, but as Voyager has done all along, we see something different!" Dodd chuckled. "It was still making discoveries."

Stone added, "It was 2012 before we actually reached the boundary, at a distance of 11 billion miles, from the sun. That's a really remarkable journey. It was something we had been planning but really couldn't commit to because we couldn't say how big the bubble was. . . . Voyager 1 has been in interstellar space since 2012. Voyager 2 is still inside but approaching the boundary in a few years."[3]

"We knew there was stuff in interstellar space," Stone continued. "There's a wind that occurred from a supernova, from five or ten million years ago, that's coming from a particular direction, and what gives the heliosphere its comet-like shape is the interstellar wind. So we knew there was something outside pushing back against the sun, we just didn't know where that would be. We crossed that boundary at 121 AU, so that's how far the boundary was." An AU is an astronomical unit, the distance from the Earth to the sun, or about 93 million miles. So 93 million miles times 121 is . . . a really big number. Voyager 1 is very, very far away, about 11.3 billion miles from Earth.

So what is next for the tenacious Voyagers? As of this writing, Voyager 1 is over 13 billion miles from Earth—it takes radio messages traveling at the speed of light almost twenty hours to reach the probe. Voyager 2 is almost 11 billion miles away, on its way toward the great crossing already accomplished by its fellow spacecraft. Within the next few years it too will be in true interstellar space.[4]

Both the Voyagers are headed toward other star systems. The Voyagers will encounter the Oort Cloud, a region of comets, ice, and other comet junk that surrounds the outer solar system, in about 300 years. It will take another 14,000 to 28,000 years to exit it completely (the extent of the Oort Cloud is unclear). The outer regions of the solar system are that far away—the far edge of the Oort Cloud is estimated to be as far as 100,000 AU, for those of you who are counting. Then, in about 40,000 years, Voyager 1 will pass within 1.6 light years of a star called Gliese 445, and, unless grabbed by some extraterrestrials during that passage, will continue on its way. At roughly the same time, Voyager 2 will pass within about 1.7 light years of the star Ross 248. The overlapping timing is completely coincidental.

But neither of the spacecraft will live to see these triumphs of the human spirit. By the mid-2020s the ability of their nuclear power supplies to generate sufficient energy to continue operations will have dwindled sufficiently that the machines will not be able to run their instruments, and their radio signals, already weak, will be nearly impossible to monitor from Earth.

Dodd explained: "By 2025, we'll have to turn off our last instrument based on power constraints. We can still send back engineering signals if we want to, but not any more science data after that. Given the strength of the Deep Space Network, we can arrange to get signals back after 2025 . . . the DSN is a finite resource, but technically, it could be done. So we might turn off the tape recorder that's keeping science data, for instance, but some other devices are being kept on not so much because they are doing important work but because they actually act as heaters and keep the spacecraft warm and operational."[5] That's right—excess heat from Voyager's instrumentation is being used to keep the spacecraft toasty inside. But eventually the Voyagers will give up the ghost, switched off by mission control. Of course, they don't have to be switched off; JPL can just let them run as long as they run. But the human impulse for control over technology—as well as a genuine affection for machines like Voyager—will probably result in a shut-down command being sent before it is too late to do so.

Barring any new and dazzling discoveries in interstellar space, the Voyagers will continue to send back routine data from the region, gradually building up an ever-expanding picture of what occurs beyond the influence of our own star, the sun. And then, sometime in the mid-2020s, they will wink out of existence so far as we on Earth are concerned, their radios silent and the giant tracking dishes on Earth focused elsewhere. The missions of the twin spacecraft will have been completed, with just under fifty years in space.

Not a bad record for a couple of repurposed Mariner spacecraft first designed in the 1960s.

Speaking of records, there is one more part of the Voyager tale that bears retelling. As with the Pioneer plaques that may one day announce our presence to extraterrestrials, so do the Voyagers carry a human story. Once again, Carl Sagan was enlisted to craft a message to be carried into the great beyond, working with a committee of NASA-selected personnel. Together they decided that this message should be something more than an informative metallic postcard, as was flown on the Pioneers. The Voyagers each carried a gold-plated metal record—that's right, a one-sided version of an old-school vinyl LP recording. The platters had imagery not dissimilar to the graphic representations on the Pioneer plaques, except that the images of humans were removed—NASA had received enough complaints about "naked people" being represented on the Pioneer plaques and did not want a repeat of that experience. But, on the recording itself, a treasure trove of human experience was preserved.

Fig. 24.2. The golden record is attached to Voyager 1. Your parents might recognize it as an LP—we can only hope that the aliens do. A stylus and photo cartridge are included for playback. Image from NASA.

The selection process of material that would best represent our planet took about a year. Natural sounds were recorded—sounds of the surf breaking on a shoreline, wind, thunder, and other earthly sounds were included. Representative samples of animal noises were inserted—whale and bird song among them—as were phrases from fifty-five languages, both modern and ancient.

Music was also in the mix, with selections from Bach, Mozart, Beethoven, and Stravinsky in the classical genre. Interestingly, the Stravinsky choice was his *Rite of Spring*, which was so alarming to the public when first performed in France in 1913 that it caused a near riot at the Paris Opera House. The *Rite* is not an easy piece of music to comprehend and appreciate, even for humans, and we must hope that any extraterrestrial listeners will not have a similar reaction—or simply find it sufficiently disturbing that they decide to drop by and squash the culture that invented it.

More contemporary tunes from Guan Pinghu, Blind Willie Johnson, Kesarbai Kerkar, Valya Balkanska, and even Azerbaijani folk music, were added. Controversially, "Johnny B. Goode," by Chuck Berry, was part of this list of recordings, with some on the team and elsewhere complaining that it was "adolescent." Sagan quipped, "There are a lot of adolescents on the planet," implying that young people should be represented for their quantity, if not for their musical taste.[6] In terms of geography, music from the Americas, Asia, Africa, Europe, and just about every other major continent and large cultural group was included.

Photographic images from Earth were also encoded—these were digitized and imprinted on the record alongside the analog recordings. Cities, landscapes, sunsets, animals, and a dizzying variety of other were all a part of the collection.

And, in the truest nod to the era, a phonographic cartridge and needle were included inside a small sealed case. The necessary turntable was not—the aliens will have to figure that out for themselves.

As Sagan said of the addition of a record to the Voyagers, "The spacecraft will be encountered and the record played only if there are advanced space-faring civilizations in interstellar space, but the launching of this bottle into the cosmic ocean says something very hopeful about life on this planet."[7] iPods be damned; advanced alien species are expected to have jukeboxes.

As the Voyagers continue to sail the reaches of the "cosmic ocean," as Sagan referred to the far reaches of space, we remain hopeful.

Chapter 25

FLASH FORWARD: GOING INTERSTELLAR

It's incredible that the Voyagers have lasted long enough to become interstellar explorers in their own right, but what about the idea of creating purpose-built spacecraft to explore nearby stars? With the explosion in exoplanet discoveries in the past two decades—now numbering over 3,500 confirmed finds, with many, many thousands more suspected—we have increasing reason to believe that some of these planets may be suitable for life. Various projects, like SETI—the Search for Extraterrestrial Intelligence—have been listening in for decades, hoping to detect a radio signal from a society orbiting a distant star, with no confirmed successes. As with our explorations of Mars, we are lowering our sights somewhat, hoping to find at least basic life—perhaps on the level of bacteria, or with luck something more complex—on a planet orbiting a nearby star. But doing that will likely involve going there. Some inferences can be made remotely—when we have large enough telescopes in space to perform detailed and routine identification of the gases in exoplanet atmospheres, for example—but more involved investigation will require a visit. How might this be accomplished?

NASA has proposed a short list of interstellar missions in the past, but the other stars are far away—very far. The nearest star to our solar system is Proxima Centauri, "just" 4.24 light years away, or over 25 trillion miles. That's a long stretch, and it takes light over four years to cross that distance. Still other stars, many with indications of more promising planetary systems, are much, much farther away.

Most traditional spacecraft designs for traveling to other planetary systems would depend on terrifically powerful rocket engines and a huge gravitational boost from the sun or a large planet to sling them out of our solar system at great speed, and it would still take many decades, or as much as a century, to reach even that closest star. No US government program has yet lasted that long—NASA itself is only sixty years old. But there are other alternatives, probes with far less mass, that might be capable of returning basic data on distant planets far more quickly. One such design would contain an electronic core about the size of a potato chip and would travel on a beam of light.

The Breakthrough Starshot initiative was started in 2016 to develop tiny interstellar probes. Prominent members of the effort include professor of astronomy Avi Loeb, of Harvard University, and Pete Worden, the former director of NASA's Ames Research Center. The late Stephen Hawking was involved before his death, and the project is being funded by Russian billionaire Yuri Milner, who committed $100 million to his broader Breakthrough Initiatives, which also includes a SETI component.[1]

In a nutshell, Breakthrough Starshot proposes that we first utilize "nanoprobes" for interstellar flight, tiny spacecraft that, when combined in swarms, can send back a large amount of data that can be put together to give us a good picture of what conditions prevail in neighboring star systems. Potentially habitable planets may exist in the Alpha Centauri system (home of Proxima Centauri), and there are many other candidates at distances ranging from six to fifty light years away.

This fleet of nanoprobes would ride to Earth orbit in a single rocket. Each nanoprobe would consist of a small computer processor, cameras, and other sensing devices, along with highly miniaturized navigation, micro-propulsion technology, and a power supply. These components would be affixed to a "sail," a lightweight structure that could range in size from a dinner plate to a large kite, and would either be a flat surface or possibly an inflatable sphere.

Thousands of these nanoprobes would make up the swarm, and once in orbit they would be propelled out of the solar system by an array of powerful lasers based either on Earth or in Earth orbit. The light from the lasers would be pointed at the swarm, and the tiny amount of pressure exerted on the nanoprobes' sails would propel them at ever increasing velocity toward the swarm's targeted star system. This would allow for an ultimate speed approaching 15 to 20 percent the speed of light, far faster than more traditional alternatives. The Starshot swarm would eventually be traveling at speeds of well over 100 million miles per hour, though even at that speed it would still take over twenty years to reach the Alpha Centauri system and another four years before Earth received the first images and science data.

Fig. 25.1. A NASA rendering of how a "lightsail" could be propelled with a laser beam from Earth. Thousands of these probes would be sent with multiple beams to a nearby star. Image from NASA.

There is a lot of engineering to be accomplished before such an endeavor can be undertaken, however. Miniaturized maneuvering thrusters must be developed and tested for use on the nanoprobes. The material for the lightsail must also be tested, to assure proper deployment in space and its ability to withstand the heat generated by the laser beam. Cameras, processor chips, batteries, communication and navigation systems, all at a scale of a gram or less, must be perfected. And, of course, the giant lasers required to push the Lilliputian spacecraft must be designed and built—and possibly launched—before the project can be initiated.

Finally, a means of slowing the nanoprobes as they near the target star still needs to be worked out. Various techniques, many deriving their functionality from the same principles that govern a sailboat, must be investigated in order to complete something more profound than a quick (very quick) flyby of the distant star system—stopping there and orbiting the local planets would be the most desirable outcome. But the scientists, engineers, and policymakers leading the high-profile team in charge of the program have not identified anything they consider to be a deal breaker, to use their own term.

In a few years, NASA will launch the James Webb Space Telescope, the first optical instrument capable of limited imaging of worlds circling other stars. It will join the recently launched TESS, the Transiting Exoplanet Survey Satellite, in the identification and assessment of nearby candidate stars and worlds that surround them. The ever lengthening list of confirmed extrasolar planets will grow quickly, and some will likely show signs of possible biological activity in their atmospheres, such as the presence of methane or other gases that can be the result of metabolism.

If the Breakthrough program succeeds, you can expect others to follow, especially if Breakthrough Starshot's laser array—likely the most expensive and complicated component of the program—is made available for use by other organizations and countries.

Chapter 26

NOT BY A LONGSHOT

Stepping back in time a bit to the 1980s, there was a precursor to Breakthrough Starshot, and it was called, appropriately enough, Project Longshot. The less-than-inspiring moniker was thought up by an unlikely combination of NASA, the US Naval Academy, and the USRA, a nonprofit research foundation founded by NASA administrator James Webb and Frederick Seitz, the president of the National Academy of Sciences, so it had a solid pedigree. But Longshot was, well, a longshot, yet it was pondered for over a decade. Longshot roughly overlapped the Strategic Defense Initiative (more popularly known as "Star Wars") era of the Reagan administration, when much seemed possible in advanced space technologies, but for the most part it ultimately proved to be impossible. Longshot in particular would have been vastly expensive and difficult to produce.

The unmanned interstellar probe was designed along a familiar theme: a single, large space probe propelled by a powerful rocket engine. But this was no ordinary chemical rocket—the spacecraft would have been propelled by a fission reactor, a miniaturized version of those used to power nuclear submarines and other naval vessels. On the Longshot spacecraft, the fission reactor would power a cluster of lasers that would fire into a chamber containing fuel in the form of helium-3/deuterium pellets that would be injected from a holding tank. With sufficient heat, these would be forced into a fusion reaction that would propel the ship at great and ever increasing speeds. Longshot would have been the first fission-fusion spacecraft in history, if it had worked. In fact, it would have been the first successful fusion reactor in history—something we are still trying to accomplish.

The giant ship would have weighed dangerously close to a million pounds—design estimates were about 873,000 pounds, but that would likely have increased—most of this mass being in the form of the fuel pellets needed for the long journey. Even with all this, though, it would have only reached speeds of about five percent the speed of light. The massive spacecraft would have been assembled at the space station the Reagan administration was planning for the mid-1980s, to be called "Freedom."

Longshot would have utilized rocket stages just as today's rockets do, but staging would have been done on a far more leisurely, interstellar time scale, jettisoning the nuclear fuel tanks at roughly thirty-year intervals (the Saturn V staged every few minutes during launch). By the time the spacecraft arrived at its stellar destination, everything not absolutely necessary—even the engines—would have been discarded. This still left a beefy instrumented probe of over thirty tons. The US Navy doesn't think small, even for space.

Communication with Earth was intended to be via laser beam, which was also powered by the fission reactor. The interstellar craft was designed to orbit the exoplanet (the term for any planet outside our own solar system) or star it was sent to, and to send back data for years via its nuclear-powered laser communication system.

Brazenly optimistic projections claimed that the four-light-year voyage to Alpha Centauri would have taken about one hundred years at just under 5 percent the speed of light, so, in comparison, the Breakthrough Initiative's nanoprobes are Ferraris next to Longshot's lumbering freight truck.

There were other hurdles that would have to be overcome to accomplish Project Longshot, beyond the enormous mass of the spacecraft. One was the need for robust artificial intelligence computer systems to control the spacecraft when it was so far away that one-way radio signals would take up to four years to reach home, and return commands another four years—not exactly joysticking distance. Recall that this was the era of IBM's original Personal Computer as well as the Apple II—not exactly Terminator-caliber AI capability.

The authors of the study did note that some technological advancements would be required—all that helium-3 would have to come from somewhere, as it is not readily available at your local Shell station, or anywhere else on Earth for that matter. Mining it from Jupiter was one suggestion put forth—not a small challenge in itself. The designers estimated that it could take twenty to thirty years to achieve the needed levels of technology, and even that turns out to be optimistic.

The command and control structure on Earth would also have to last well over one hundred years to see Longshot arrive at its destination, a timespan not yet achieved in American history for any single technological project (though the Voyager project will be coming up on its fiftieth anniversary in about eight years).

Project Longshot ultimately never moved beyond a study, and it has been characterized as a "gutsy academic exercise"—surely an accurate assessment.[1] The Breakthrough Starshot interstellar nanoprobes are a better, faster, and far more achievable alternative.

Chapter 27

RUDDY CHUNKS OF A RED PLANET: MARS 5M

The Soviet Union's efforts to explore Mars, which have been followed by those of the Russian Federation, have not been very successful, but it certainly has not been for a lack of effort or ingenuity. Their Mars program has been filled with innovative and daring machines designed to orbit the planet, land on its surface, and even traverse it with tethered mini-rovers that were flown before the Viking landers set down on the planet. In fact, the first machine to successfully navigate to the Martian surface was the Soviet Mars 3, which headed off toward the red planet in 1971 along with its twin spacecraft, Mars 2. Both spacecraft were comprised of an orbiter and lander, and successfully entered Martian orbit in November (Mars 2) and December (Mars 3) of 1971.

These Mars orbiter/lander combinations were in the same mass ballpark as the Vikings, at just under 6,000 pounds each. The orbiters were outfitted to study Martian topography, its atmosphere, and the radiation levels near the planet for a period of months. The landers were even more ambitious—they were roughly spherical with a flattened bottom and piston-driven petals not unlike the AVCO design for the Voyager Mars lander. When the probe landed, the petals would open, forcing the probe upright. Onboard each lander were two television cameras; a mass spectrometer; temperature, wind, and pressure sensors; and various instruments to measure soil dynamics and composition. There was also an arm-mounted scoop intended to deliver soil samples to the instrumentation aboard to allow for basic soil analysis.

Fig. 27.1. Mars 2 was the first successful Soviet orbiter. Due to a lack of reprogrammability, the Mars 2 and Mars 3 orbiters began to photograph the planet automatically upon their arrival, while it was covered by a titanic dust storm—it looked like a ruddy billiard ball in space. Later images revealed more detail. Image from NASA/NSSDC.

Perhaps most remarkable for the time was a small rover that was carried inside the spacecraft. Called Prop-M, the tiny ten-pound machine looked like a tall metal pizza box and was connected to the lander via a fifty-foot cable. Instead of wheels, it had two articulated skis, one on each side, that would allow it to slowly inch forward. There were bump-bars on the front to provide simple obstacle avoidance. Instrumentation was basic, but it included a penetrometer—a device that would contact the soil and provide information on density—as well as a radiation detector.

When the twin spacecraft arrived at Mars, the United States'

Mariner 9 orbiter had already been operating there for a few weeks and had sent home the bad news: a global dust storm had blanketed the planet, completely obscuring the view from orbit. JPL was able to reprogram Mariner 9 to wait out the storm and resume operations once the dust had settled, but the Soviet orbiters were not reprogrammable from the ground, and they performed their duties automatically, sending back about sixty images, mostly of a featureless, dust-covered planet. The mission was a success operationally but provided little visual data that was useful. Other measurements—those that did not rely on visual reconnaissance of the planet—were more successful. Data regarding the atmospheric density, composition, and temperature of the Martian atmosphere were gathered, as was information about the weak magnetic field and the gravitational field. The orbiters successfully returned data for three months, and missed imaging through a dust-free atmosphere below by just a few weeks.

Both spacecraft also dispatched their landers to the Martian surface automatically. Down went the Mars 2 lander, firing its rockets to initiate a deorbit and to slow the descent. But the descent system malfunctioned, the parachute did not open, and Mars 2 slammed into the planet. It was, however, the first spacecraft to reach the surface of Mars, albeit via high speed impact.

Mars 3 fared better, with the parachute opening as designed and setting the machine safely onto the surface on December 2, 1971. The petals enclosing the lander deployed properly, righting it and preparing the probe for surface operations. After about ninety seconds, the TV cameras began transmitting—and then abruptly stopped. Soviet engineers scrambled to fix the problem but to no avail. For reasons still not completely understood, the onboard electronics seemed to have failed. A partial image of seventy lines of video was returned, showing what looked like a horizon topped by lines of static, but according to subsequent analysis, nothing in the image was recognizable as a landform. It must have been crushing to experience such immediate success followed by total failure of the machinery upon touchdown.

One possible cause of the failure was offered—the landers had both descended into the fierce dust storm, and these storms are thought to generate powerful electrical discharges. It is possible that a huge electrical jolt hit the lander before or after it set down, disabling the circuitry onboard. In any case, while it represented another huge step in planetary exploration—the first soft landing on Mars—the Mars 3 lander was not successful in completing its mission.

Another issue was that the engineers had been rushed to meet a launch date before they were ready. Soviet Mars flybys scheduled for 1969, in preparation for the Mars 2 and 3 orbiter and lander combinations, had both failed during launch. In light of this and other factors dogging the Mars 2 and 3 program, the engineers advised a delay from the 1971 launch window to a 1973 date. (Mars is in favorable position for launches from Earth every two years.) But with the American Mariner 9 orbiter planned for 1971, Soviet politicians would not allow the delay, and the probes were dispatched on the originally planned date. A lack of preparation time appears to have added to the program's woes.

Regardless, the Soviet Union's planetary exploration program marched on. This was the space race, and the manned lunar landings had already been ceded to the Americans—Soviet efforts had been doomed by the failure of the Russian answer to NASA's Saturn V, the giant N1 moon rocket. Russia therefore hoped to achieve glory via robotic exploration. No less than four additional Soviet spacecraft were flung toward Mars during the 1973 opportunity, with mixed results. Mars 4 and 5 were orbiters, intended to return imagery and other data from high above the planet. Unfortunately, a single transistor failure appears to have doomed Mars 4, and it sailed past the planet without firing its braking rockets. Mars 5 was more successful, entering Martian orbit in February 1974, but the pressure hull containing the electronics developed a leak and the probe failed by its twenty-third orbit. Despite the electronic issues, both spacecraft managed to return useful images this time.

Mars 6 and 7 were landers without orbiters—they did include a "space only" coast stage that would fly by the planet, but the role of orbiting relay stations was to have been taken by Mars 4 and 5. Mars 6 and 7 both launched successfully in August 1973. The massive machines were about 7,500 pounds each, or about the same mass as the combined orbiter/lander of Mars 2 and 3. Each lander contained a mass spectrometer and thermometers and barometers to measure atmospheric conditions. Other instrumentation was included in the coast stage to measure environmental conditions in near-Martian space.

Mars 6 successfully navigated into the Martian atmosphere and began the execution of its descent sequence, deploying its parachute, but just as the braking rockets were about to fire, contact was lost. Mars 7 separated from its coast stage but failed to fire its retrorockets to orient itself for a descent into the Martian atmosphere and flew past the planet, missing it by about 800 miles. You can see a pattern forming here . . .

It should be noted that the operations required to perform these landings were once again completely automated, based on navigational prompts gained from optical star trackers and a Mars tracker. Little could be done from the ground, and if something went wrong in flight it was essentially a fait accompli—the landing would not succeed. While the engineering involved was quite ingenious in its simplicity, the sequence of steps was not simple, involving the firing of small rockets to slow the craft and spin the lander for stability during entry, and these may have contributed to the mission's failure. But you've got to give the USSR points for a clever effort.

Stubbornly steadfast, the Soviets hatched plans to up the stakes once again—rather than repeat the same effort, why not aim higher to outdo the Americans? A far grander attempt was contemplated for later in the 1970s, a mission profile that still eludes the spacefaring powers—bringing back a sample from the Martian surface. If successful, this great leap in robotic science at Mars would really put the Soviet Union on the map in planetary exploration, leaving the United States in the dust, so to speak.

This mission, called Mars 5M, evolved out of previous sample-return plans that depended on the unfortunate N1 rocket the Soviets had been developing without success. This earlier iteration weighed in at an impressive ninety-eight tons—an unprecedented mass for an unmanned spacecraft—this was in the same neighborhood as the combined Apollo Command/Service Module and Lunar Module. However, since the N1 rocket had failed in every launch attempt (four in all), those missions were canceled, and out of that came the leaner Mars 5M.

The 5M mission was still a whopper, however. Three spacecraft components would have been launched on separate Proton boosters and would dock in Earth orbit, assembling themselves for interplanetary flight robotically. This undertaking was ultimately deemed to be too complex, and the mission was scaled down. In its final iteration, two spacecraft would be launched on two Proton boosters. By the mid-1970s, the Proton could claim a launch success rate of 90 percent, which must have been a stunning relief after the fiasco of the N1 booster (which had a failure rate of 100 percent). The Proton was capable of lifting 50,000 pounds to low Earth orbit, but two would still be required for this complicated and audacious mission.

Fig. 27.2. The Mars 5M sample-return lander. The inverted umbrella structure at the bottom would have replaced a parachute to slow the massive machine's descent. Image from NASA.

The lander alone would still be a porker at almost 16,600 pounds—far heavier than any previous robotic craft. Even NASA's massive Cassini Saturn orbiter, which would not launch until 1997, was "only" 12,600 pounds, so in terms of robotic spacecraft 5M was a big one. One launch would carry a fueled upper stage, complete with a docking mechanism, and the second would carry the actual lander/sample-return spacecraft and an orbiting component with another fueled return rocket stage. After linking-up up in Earth orbit, the combined spacecraft would head off to Mars.

After a cruise of approximately eleven months, the orbiter would separate to take up position around Mars, and the lander would continue to a direct entry into the Martian atmosphere. As it neared the surface, the lander would deploy an "inverse umbrella" of beryllium spokes with fiberglass cloth, which would expand out to thirty-six feet, providing a large drag area to slow the craft, and would then use braking rockets to decelerate to its final soft landing—no parachutes required.

After landing, the surrounding terrain would be surveyed by transmitting imagery back to Earth, and a suitable sampling area would be selected. Samples would be collected and dumped into the return stage, which would then prepare to be sent back to Earth. One complication was that the position of the lander would need to be known with high precision to calculate the proper trajectory back to Earth—a complex problem even with today's technology, and extremely optimistic with that of the 1970s. Various means of obtaining samples were considered, including a scoop and possibly a drill—something not attempted by NASA until the twenty-first century with the Curiosity rover.

The Mars ascent stage would then be launched to Martian orbit, where it would link up with the waiting sample-return vehicle. Once the sample was inside the latter, it would be sterilized in place, using high heat to assure that no nasty Mars bugs would be unleashed upon an unsuspecting populace back home—while few thought the sample would be particularly dangerous, visions of poisonous red weeds or noxious killer microbes smothering our home planet were unacceptable to Soviet scientists, so precautions were in order. Dark visions of interplanetary invasion die hard in the human imagination, and rightly so—we just do not know much about what might dwell on the surface of other planets.

At one point, a rendezvous with an orbiting Soviet space station was considered—the cosmonauts onboard would perform the first analysis of the sample before deciding if it was safe to return to Earth. Ultimately, however, it was decided that the sample-return vessel would return to Earth in a fashion as unique as the umbrella-clad Mars lander. Instead of using a parachute, since weight was a still huge issue at the end of the mission, the return vessel would simply plunge into the Earth's atmosphere at 27,000 miles per hour, be slowed by its heat shield, and then slam into the hard surface in the sparsely populated eastern end of the Soviet empire. The sample's container was designed to withstand the harsh landing profile, and it would likely have done so. The plan was for it to carry a small radioactive element aboard that would aid in its detection from the air, and the landing zone was thought to be controllable within about a twenty-four-mile range. Simple . . . in theory anyway.

Design, construction, and testing were still underway in 1978 when the project was ultimately canceled. Reasons included concerns about its complexity and also the fact that this Mars mission was competing with Soviet lunar sample-return projects and Veneras 11 and 12 (the sample-return components of these missions were also ultimately canceled for similar reasons). There were also political issues that may have contributed—the Soviet planetary programs were executed by competing design bureaus that were not unlike NASA field centers; however, in the case of the Soviet agencies, the competition was far fiercer and the leadership of the planetary program was less centralized than it was in the United States under NASA.

Despite the cancellation of the project, low-level efforts continued, as will happen in scientific and engineering establishments staffed by devoted individuals. A Russian Mars sample-return mission has been studied in at least three more iterations, one during the waning years of the Soviet Union and two or three more since the birth of the

Russian Federation. As of 2012, plans were in place to attempt such a mission in the early 2020s, but even with modern technology this is a complex undertaking. Other Russian Mars programs have met with mixed results—most recently the failed landing attempt of the ExoMars lander in 2016 (launched jointly with the European Space Agency [ESA]), when the spacecraft miscalculated its altitude, separated from its parachute early, and fired its landing rockets for only a tenth of the time necessary to set down successfully. Sample-return missions are much more complex, and will have to wait.

Had any version of the Mars 5M program succeeded, it would have been a triumph for Soviet space science, and made a huge contribution to planetary science in general. It would also have been a marvelous recoupment of national self-esteem, coming about a decade after the loss of the race to land humans on the moon. The Soviet Union did manage to return samples from the moon with a robotic probe in 1970, but a Mars sample-return mission was a far more challenging undertaking. Even NASA, which has had great success at the robotic exploration of the red planet, has been slow to prepare for a Martian sample return.

In time, one of the spacefaring nations will attempt this mission—other than a manned landing on the planet, a sample return is the holy grail of Mars exploration. This may be achieved by the United States, or perhaps by China, which has clear ambitions to explore Mars; or it may ultimately be accomplished by private industry—perhaps Elon Musk's behemoth Big Falcon Rocket (BFR) will scoop up some Martian rocks before a return trip home, when it makes its planned flight to Mars in the 2020s.

For now, however, whatever remaining secrets hide within Martian rocks and soil remain just that—secrets. For the foreseeable future, Mars will be studied by machines that analyze the planet in place with instruments they carry onboard. We will learn more when NASA's Mars 2020 rover lands there in a few years, along with Russia and ESA's ExoMars rover in the same time frame, but until then we must be content with studying data from NASA's Curiosity rover and from the small fleet of orbiters that ply their silent paths around the planet.

CHAPTER 28

FLASH FORWARD: ROBOTIC RENDEZVOUS AT MARS

The possibility of a Mars sample-return mission has been coveted by planetary scientists for decades, but sample retrieval, rendezvous, docking, and sample transfer are difficult to accomplish robotically, as we saw in the previous chapter. While much work has been done toward this goal, no mission has yet been launched to accomplish a return of Martian soil. However, a number of small projects to prepare for this eventuality are proceeding at various NASA field centers and elsewhere, and one of the more interesting components involves how to simplify the robotic rendezvous and docking of the Martian samples with an orbiting spacecraft to return them to Earth from Martian orbit. JPL is developing a new technique that may greatly simplify the process using superconductors.

Utilizing magnets to passively attract two spacecraft in proximity to each other is not a new idea, and sounds remarkably simple, yet it takes some very advanced technology to accomplish. A new technique called magnetic flux pinning, first investigated at Cornell University, updates this concept by using superconducting magnets to assist with rendezvous and docking.

Traditional sample-return designs use robotic arms to grapple sample containers in Martian orbit, or more traditional docking probes to facilitate the joining of two spacecraft. These systems require complex active control from Earth, especially to close the last few inches between the two craft and make contact with the sample container, which may be tumbling or spinning in orbit. The new concept, called magnetic-flux-pinning rendezvous, uses superconducting physics and magnets to bring two spacecraft together passively. Unlike magnets alone, magnetic flux pinning can bring the two spacecraft to a stable proximity relative to each other without active position control or contact between the spacecraft—once the orbiting sample container gets close to the Earth return vehicle, it just stops and hangs there, with any motions instantly arrested by magnetic fields.

A Mars sample-return mission using magnetic flux pinning might look like this: sometime in the 2020s, a robotically controlled rocket ascends from the Martian surface with samples inside. This rocket has permanent magnets embedded in its forward structure where the samples are located. An Earth Return Vehicle, which has subsequently been sent into Martian orbit, contains a set of superconductors and a set of permanent magnets that match the magnets on the orbiting sample. When it is time for the two to rendezvous, the superconductors on the Earth Return Vehicle are chilled to induce a superconducting state, then permanent magnets matching those embedded in the orbiting sample craft are used to "train" the superconductor. These training magnets are then retracted, clearing the superconductors for docking. As the orbiting sample vehicle nears the Earth Return Vehicle, the superconductor "remembers" the magnetic flux signature and draws in the permanent magnets on the orbiting sample from up to about eighteen inches away. The two spacecraft are then able to rendezvous safely, with the orbiting sample being magnetically held at a slight standoff. Even if the sample container is tumbling in space, the superconductors would act to drive it into the proper configuration without any mechanical contact. The sample container can then be cleaned before being brought inside the Earth Return Vehicle for return to Earth, reducing contamination concerns. It sounds complex, but in would be quite simple in operation. A key advantage is that such a system removed many of the concerns about the long delay times of radio signals traveling from Earth to Mars, which have been used to accomplish more traditional rendezvous in space.

A sample-return mission has not yet been fully designed, nor has one been funded by NASA, I'll remind you—complain to your senator if you don't like it.

Besides reducing complexity, the magnetic-flux-pinning technique may help to alleviate planetary protection concerns—by transferring the samples en masse, and cleaning them with chemicals or a laser before return to Earth, the potential chain of contamination is broken. Risks to Earth therefore are minimized—nobody in a healthy state of mind wants clusters of Martian tentacles encircling our planet, and hopefully that includes you.

Chapter 29

JUPITER'S REVENGE

If you walk into a rainstorm with a half-opened umbrella, it won't work very well. You will get wet, and it will have failed in its mission to keep you dry. If you are using an older umbrella that's been stored in a drafty garbage for a few years, that might explain the dysfunction. This is pretty much the story of the Galileo mission to Jupiter. The magnificent Jupiter orbiter had its own malfunctioning umbrella, and the story of how that occurred—and its long and tortured path to launch—is as fascinating as the mission itself.

The short version is that the Galileo probe was delivered to Earth orbit by the space shuttle *Atlantis* in 1989, and arrived at Jupiter—after a series of gravity assists provided by flying past both Venus and Earth to pick up speed—in December 1995. In 2003, after fourteen years of yeoman service, the spacecraft was sent plunging into Jupiter to avoid it crashing into one of the Jovian moons—nobody wanted to inadvertently contaminate these promising places, only to later discover that extraterrestrial life had been wiped out by errant flu germs or some other microbial hitchhikers carried on a crashed NASA spacecraft.

But the long journey of Galileo started much earlier than that 1989 launch. Jupiter had been identified as a top priority for planetary exploration in a 1968 NASA study called the Planetary Science Decadal Survey,[1] part of a series of once-every-ten-year studies to determine the United States' priorities in robotic exploration. The Galileo project was officially initiated at JPL in 1977, even as the Voyagers were preparing to launch, bound for the outer solar system. Galileo's mission planners would have time to adjust the parameters for the mission as new data came back from the Voyagers—as it turned out, they would have almost a decade, though that was not the original plan.

The first date considered for a launch to Jupiter was in 1982 via the shuttle *Columbia*. At the time, NASA was specifying that all such launches had to be accomplished via the shuttle—there were concerns that using conventional unmanned rockets to send robotic craft (and other payloads) into space would pull too much business away from the shuttle, lessening its viability. But the shuttle was still new—it had originally been slated to fly in the late 1970s, but the first launch slipped to 1981—and delays in the program kept pushing back the launch date for Galileo, first to 1984, then 1986.

The delay was welcomed by some of the engineers—it gave JPL more time to work out any bugs in the spacecraft. They were also working on another factor: how to get Galileo out of Earth orbit and off to Jupiter. This requires a lot of energy, and the plan was to use an upper stage called Centaur, designed years before by the Air Force, to do so. The Centaur was a powerful liquid-fueled rocket that usually rode into space atop an Atlas or Titan rocket, both of which were unmanned. In this case, however, the plan involved placing a fully fueled Centaur in the payload bay of the shuttle, with Galileo attached to it—the combined spacecraft would be released from the shuttle, nudged away from it, and then ignite and speed off to Jupiter. Easy peasy.

Then came the *Challenger* accident.

In January 1986, the shuttle *Challenger* exploded on its way into orbit. This story is well known—the launch occurred on an extraordinarily cold January morning, and the rubber O-rings used to seal segments of *Challenger*'s solid rocket boosters were not up to the job—they shrank and got hard and brittle in the cold, and hot gases blew past them, torching the sides of the shuttle's large fuel tank. This resulted in a catastrophic breakup of the shuttle as it ascended, killing the crew. The shuttle program was shut down for investigation, which included a reevaluation of NASA's plans to use the shuttle for all its launches, many of which could be accomplished with much cheaper, expendable rockets like the Atlas, Delta, or Titan.

Then the committees looking into the disaster realized that NASA was also planning to launch a fully fueled Centaur stage in the payload bay of the shuttle. An already dangerous spacecraft system would be carrying a 48,000-pound bomb filled with explosive liquid hydrogen and liquid oxygen inside its cargo area. The Centaur would need the cryogenic fuels to be maintained and vented as they boiled off right up until deployment. Further, the Centaur had never been "man-rated," meaning that it was not designed with human safety in mind. It was a thin-skinned, lightweight rocket stage that could puncture or leak easily. John Young, a cucumber-cool Apollo and shuttle astronaut, who was as far from alarmist as an astronaut could get, took to calling it the "Death Star," as he felt the Centaur was far too dangerous to be used in this application.[2] There was also the possibility that, if the shuttle suffered a mishap during ascent, it would have to perform an emergency landing with the heavy Centaur stage in its payload bay, and with the explosive fuels sloshing around—a prescription for possible disaster. Among other concerns was the fact that the shuttle's landing gear had never been designed to cope with such beefy loads during touchdown. In hindsight, the plan was bold, but brash, and probably ill-advised.

It's worth noting that the Centaur had enjoyed a far better than average success rate since its first flights in 1962—by the 1980s, only two out of thirty-five Centaur flights had resulted in complete failures, though it experienced a number of smaller problems. Nevertheless, just days before the *Challenger* accident, astronaut Rick Hauck, who was to command a mission that would shortly follow the launch of *Challenger*, approached his crew to discuss the upcoming flight. It would be the first use of a Centaur stage on the space shuttle, to launch a robotic solar probe called Ulysses. The essence of his comments, as he recalls them, were this:

> NASA is doing business different from the way it has in the past. Safety is being compromised, and if any of you want to take yourselves off this flight, I will support you.[3]

These were daunting words coming from a mission commander in the usually gung-ho astronaut team. But it had become clear in the first few years of sustained shuttle operations that overall flight procedures had been sped up and rules relaxed to accommodate an accelerated schedule, and some of the astronauts—and many engineers and managers—were getting nervous. The shuttle, which often flew close to design and safety limits, was being pushed too hard. Flying the Centaur stage could be one step too far.

After the *Challenger* explosion, the idea of flying the twenty-five-ton Centaur in the shuttle was dropped. Payloads that were not specific to a shuttle mission would fly on other rockets used by both NASA and the Air Force, and there were a number of them, all supplied by private aerospace contractors.

But this did not let Galileo off the hook—since it was a NASA science project, and had already been slated for a shuttle launch, it would await a slot on the shuttle. There was an alternative to using the Centaur, Boeing's Inertial Upper Stage, or IUS. This was a smaller, but arguably safer, solid-fueled rocket stage. The upside of the IUS was that it was a big rocket that was filled with solid fuel that only needed to be drifted free of the shuttle and lit—and off it would go. No sloshing, hard-to-maintain liquid fuels, and it was much less complex in operation than the Centaur—the IUS was far simpler to handle, load, and use.

The downsides were that it was still a big bomb in the payload bay, and also it was much less powerful than the larger Centaur. Being a solid rocket, it was also not throttleable—any course corrections and speed changes would have to be performed after it had done its job. Since the IUS was less powerful, the trip to Jupiter would also be lengthened from about two years to six—because of the less powerful rocket, Galileo would require a number of trips past the inner planets to build up enough velocity to sling it out toward Jupiter. Finally, the firing of a solid-fueled rocket also subjected its payload to a much more severe shock when starting—it was an all-or-nothing *bang* when it ignited. Even with these considerations in mind, NASA decided that the shuttle was still the best option for launching Galileo. They just wanted it to *fly*.

John Casani was one of the program managers on Galileo (I told you he was a legend in the space exploration community) and was becoming increasingly frustrated by NASA's inability to get Galileo launched. "It was really a difficult mission in that it got delayed five or six times due to scheduling problems and development problems with the shuttle. And every time it got delayed, our launch date got moved out. When the launch date gets moved, out there are cost impacts."[4] The probe cost money just sitting there, and he wanted it to be burning money in orbit around Jupiter, doing great science, not while sitting in storage on Earth.

As these decisions were being sweated out, Galileo languished for almost three years. Work continued on mission planning, but the spacecraft had been on schedule for the 1986 launch date and was essentially done. It sat in storage for the duration, with occasional inspections and bits of maintenance work. But a spacecraft like Galileo is extremely complex, made up of hundreds of thousands of parts, large and small, and it is not designed to be disassembled and serviced while sitting around for three years. Nor was Galileo designed to be stored for so long in Earth's atmosphere—the people who built it reasonably assumed that it would spend most of its time in space, transiting from Earth to Jupiter, not sitting in the oxidizing air of our planet. There were also plenty of moving parts, most coated with lubricants that could become less effective with age, and, so far as we know today, this is what ultimately came back to bite the spacecraft designers in their collective hind quarters. The "dry lubricant" used degraded quickly in Earth's atmosphere.

While most of the spacecraft ultimately worked as planned, there was one major exception: the main antenna. Up until this time, JPL and NASA had flown spacecraft with the antenna's diameter constrained by the size of the rocket launching it. The width of the nose cone, or fairing, dictated how big the antenna could be—they were one-piece, rigid dishes. Until now, this had been sufficient for planetary missions—spacecraft like the Mariners and the Voyagers could accomplish their work within this constraint (the Voyager antenna was fourteen feet in diameter). And, in fact, the original intention was to use the same dish the Voyagers used on Galileo. But Galileo would have lots more information to pass along—the instrumentation would be collecting vast amounts of data as it gathered information about Jupiter and its moons, and a larger radio dish was necessary to get all this information back to Earth. So the engineers arrived at the obvious solution—make the dish bigger, in this case sixteen feet across.

How to handle a larger radio dish was the challenge before them—spacecraft radio antennas had always been solid dishes, and they had to be small enough to fit inside the launch vehicle. The designers would need to find a way to launch a larger antenna that could travel to space in a stowed configuration, that could withstand the stresses of launch, and, due to the cancellation of the Centaur stage, also survive the departure from Earth orbit via the kick from a solid-fueled rocket. Furthermore, this antenna would need to be resilient in the face of the dramatic temperature swings from hot to cold during spaceflight and still open reliably—there would be only one chance for this antenna deployment to work.

The design the engineers chose was a ribbed umbrella configuration with a thin mesh that stretched between the ribs. The antenna was engineered to work in X band, the preferred frequency for high-speed data transmission. It would be capable of sending back 134,400 bits per second, a generous increase over the Voyager's antennas, and with enough bandwidth to transmit a frame of video data once per minute.

Such an antenna was already being built for another spacecraft, a network of Earth-orbiting relay satellites called TDRSS, for Tracking and Data Relay Satellite System. It seemed to be exactly what Galileo needed—an umbrella-like folding arrangement that would launch compactly stowed, and open once in space. The only major change is that Galileo's antenna would need to be larger, so the design was modified to accommodate this need. This would give them the boost in bandwidth that they needed for Galileo's rich return of data while still being stowable for launch. JPL took that design and recreated it in-house to their own more rigid specifications. The upside? A foldable antenna that would offer far higher data transmission rates than a rigid dish. The downside? It was a single point-of-failure—if it for some reason did not open properly, the mission could fail—and NASA abhors single points-of-failure. But it was decided to move ahead regardless.

Once the engineers were told that Galileo would be launched using the lower-powered IUS rocket, however, they realized that a deployed antenna might not do well in the environments it would encounter during its multiple swings past Venus and Earth—flying that close to the sun could cause parts of the antenna to deform or melt. A sunshade was put in place to protect it until the spacecraft was clear of the inner solar system. During the early phases of the flight, when the antenna would be released but not yet fully expanded, low-gain antennas would be used to communicate with the spacecraft instead—smaller, simpler, and more robust for the flight past Venus, but designed only for basic chores such as receiving and transmitting basic health information of the spacecraft, and receiving and acknowledging commands sent to it from the ground. Two of these small antennas were affixed to Galileo—one on a boom, and another atop the main unfolding antenna.

Are you getting that tingly uh-oh feeling yet? There's more.

The umbrella-like main antenna had eighteen ribs made from beryllium, and when folded the ribs were restrained by a set of pins midway along each rib that attached to the "tower" at the center of the antenna—and herein lay the rub, so to speak.

These pins were lubricated so that when the actuator motors on the ribs were powered up, the pins would pop free, releasing the ribs, which would pop free, allowing the umbrella to fully open—it's all simpler than it sounds. Had it worked as planned, Galileo would have been on its way to accomplish great things at Jupiter, things only hinted at during the fast flyby of the Voyagers.

If you didn't have that tingly feeling before, you should now.

It's not entirely clear what actually happened, but with some "CSI: Kennedy Space Center"–level sleuthing, the engineers think they have a pretty good idea. First of all, the lubricant was not designed to sit around for three years prior to launch, especially not in earthly environments where a certain amount of dust was sure to get inside the storage wrap.

Second, with the stand-down of the shuttle program and the ensuing restructuring of America's space efforts, the antenna assembly made cross-country trips from California to Florida and back, moving from testing in California to preparation for launch in Florida then into storage in California. These trips had been intended to be made by plane, but were ultimately carried out by flatbed truck. To save money. Yes, NASA's $1.4 billion Jupiter spectacular would save money by dragging the prized space probe's delicate components across thousands of miles of bumpy, rutted, and potholed streets, highways, and freeways from one coast to the other and back. And, as the trucks slammed and bumped their way cross-country for a total of two round trips, totaling well over 10,000 miles, those little retaining pins were sliding microscopically against the lubricant, wearing it away one tooth-rattling mile after another.

They should have called Federal Express instead.

By the time Galileo had made its final crossing to KSC (the Kennedy Space Center) for launch from pad 39, the lubricant on those pins was probably largely scraped off. Just like those aging knee joints in middle-aged athletes, the interface between the metal parts, critical to a smooth release, was essentially unlubricated, or close enough that the remaining lubricant made little difference. And lest you think there is a bit of excessive guesswork at play here, in the post-mortem investigation of why the antenna didn't open all the way, it became clear that the three ribs that did not deploy properly were those that would have vibrated the most when being transported in the rattling flatbed trailers, the motions grinding the lubricated parts against their fittings for at least a couple hundred hours. It was also noted that the "dry lubricant" used on the moving parts in the antenna—molybdenum disulfide, for those of you who will be participating in the upcoming trivia exam—ages poorly in Earth's atmosphere, and would probably not have been selected for use on Galileo had anyone known it would be sitting around in storage for a decade . . .

Oops.

But none of this was known at the time, and this part of the antenna was apparently never thoroughly inspected prior to flight. It had been sealed, so what problems could there be?

On October 18, 1989, the space shuttle *Atlantis* roared off the pad at KSC. About six hours after launch, eager to perform their primary mission, the astronauts aboard rotated the deployment mechanism that held Galileo, sitting atop the IUS stage, into position. The assembly was released, and Galileo was kicked free, drifting to a sufficient distance for the IUS solid rocket motor to fire and send it on its way. Everything seemed to be going swimmingly, and the crew moved on to other duties. Galileo was headed toward a rendezvous with Venus, then it would swing past Earth—not once, but twice—to pick up enough speed to make its way to Jupiter.

On April 11, 1991, Galileo was finally headed to the outer solar system after passing Venus. Now that it was through the hottest part of its journey, it was time to open the high-gain antenna to get the spacecraft ready for its long voyage to Jupiter. The engineers sent up the command and waited to see the confirming telemetry on their screens. They waited some more, then some more. The data indicated that the actuator motors had functioned for eight minutes, but had immediately begun to draw more power than they should have—something appeared to have stopped them from driving the antenna to full deployment. More alarming, it was soon clear that the Deep Space Network—NASA's globe-girdling system of radio dishes used for communicating with its spacecraft—was not receiving the strong signal that it should be.

JPL quickly assembled a team of over one hundred engineers and other specialists to "work the problem." They had their telemetry in hand, and a backup Galileo antenna assembly at JPL that they could experiment with to see what configuration or malfunction would cause the sensors to send similar signals.

Their diagnosis? Stuck antenna ribs. Their recommendation? Barbecue, baby.

Over a number of weeks, flight controllers rotated the spacecraft this way and that, pausing long enough to let one side chill in the deep shadow of space while the other side roasted in the sun—at about 250 degrees Fahrenheit on the sunlit side and less than-250 on the shaded side. It was hoped that the 500 degree differential would cause sufficient heating (expansion) and cooling (contraction) of the pins and their fittings to spring free, but it did not.

Fig. 29.1. Artist's impression of Galileo swinging past Jupiter. The partially deployed antenna is seen on the left. Nobody knows exactly how far it actually unfurled, but this is NASA's best guess. Image from NASA.

Their next step was to "hammer" the rib actuators—switch the motors on and off rapidly, in the hope that this might jar the pins loose. Over the course of five to six weeks, they repeated this procedure a staggering 13,000 times, but to no avail. The antenna remained stuck. Controllers even tried the hot/cold spinning the spacecraft and hammering at the same time, with the same result—nil.

JPL would have no choice but to run the mission from the smaller low-gain antennas, and the resulting data transmission would be more like a dribble through a soda straw instead of a torrent through a fire hose—just a trickle of information. The low-gain antenna, rather than using a tightly focused beam, sent out a generally spherical signal and was capable of receiving from nearly any direction as well. This is perfect if your spacecraft is off axis, with the main dish pointed away from Earth, and you just want to tell it to re-aim the high-gain dish back at home base—just a few lines of digital instructions. But as a primary communication link for all that scientific data and bandwidth-hogging imagery, it would be miserable.

This was, however, all that was available, so the next step was to deal with the problem as best they could on the Earth side of the equation. New computer programming was written to optimize Galileo's science observation time, and then to save everything to the onboard data recorder, which would later s-l-o-w-l-y send the data back to Earth. Unfortunately, the data recorder had been designed and built in the late 1970s, a decade before the delayed launch. The "hard drive" was actually a tape recorder, not dissimilar to the units on the Voyagers, and was very slow. To make things more challenging, the machine jammed in October 1995, just as Galileo was closing on Jupiter—and stayed frozen for weeks. Fortunately, after a nail-biting interval, flight engineers successfully brought the machine back to life.

Other engineers figured out a new set of compression protocols to speed the transmission of mission data through the low-gain antenna, as well as new software to optimize its operation. To give you an idea of the scale that would be involved: the high-gain antenna would have transmitted data at the aforementioned 134,400 bits per second, and the low-gain antenna could accommodate about 8–10 bits per second—about 1/13,000th of the previously anticipated data rate. It would be like going from our current broadband world, where websites download in seconds, back to a 1,200-baud dial-up modem—true misery for those of us who remember those days.

The engineers also rejiggered the Deep Space Network's reception equipment on Earth, emphasizing sensitivity to the S-band signal used by the low-gain antennas instead of the X-band signal the ground stations had been optimized for. The changes in compression protocols rejected less important data in images and science returns, while retaining the more important ones, a technique not unlike JPEG image compression used in photographs sent over the internet. While this resulted in a lot of image compression, this was a huge improvement over where they started.

The science teams did their part too, reorganizing their campaigns to emphasize the most important science returns while minimizing or delaying others.

Finally, the hardware on Earth that was intended to receive the signals from Galileo was optimized for the new low-data mission profile. From the distance of Jupiter, the signal from the low-gain antennas was less than the background static at times—truly Herculean efforts would be required to separate the wheat from the chaff, as it were. It would be like listening for a dripping faucet in a rainstorm. When possible, groups of Earth-based receiving antennas would be "arrayed" or utilized in clusters, in effect making a number of smaller receiving dishes act like one large one. As the Earth rotated, more than one of the designated receivers would be pointing in the proper direction to receive Galileo's signals, and at times up to five dishes would be online at once. Galileo was reprogrammed to send the bulk of its data during these periods.

By the time they were done, the mission planners and engineers managed to save about 70 percent of the science originally intended for the mission, an amazing achievement considering how badly compromised the spacecraft's transmitting system was. The trick was that the data transmissions from the spacecraft would have to be "throttled," depending on how many ground-based receivers were pointed its way at any given time. Galileo would be told when to increase or decrease its transmitted data rate at specific time windows. Fortunately, since Galileo recorded the information from its instruments onto its onboard tape recorder, anything missed during a downlink session could be rescheduled for later playback, assuming the data recorder continued to work properly.

As Galileo made its trek through space, characteristics of the interplanetary medium were measured. Venus was studied as Galileo made its pass to pick up speed toward the outer solar system.

Galileo's second flyby of Earth, again to pick up speed, was a squeaker. It passed our planet at an altitude of just 188 miles, which was pretty daring given that it was already traveling extremely fast and also carried a radioactive power supply—nobody wanted a miscalculation in trajectory to result in a breakup of the spacecraft in the atmosphere and toxic plutonium dust scattered over multiple continents. NASA posited that the efforts of such a release of plutonium would be minimal, yet some independent scientists disagreed, suggesting that large increases in diseases such as lung cancer could have been a result. The truth is, nobody can really be sure what the effects of a chunk of plutonium vaporizing in the upper atmosphere would be—there are simply too many variables.[5]

Galileo shot past Earth and headed toward the asteroid belt, which all outer-planet probes must traverse. The belt is not as scary as it looks in the movies though—in most areas you can't even see one asteroid from another. But it still required some planning, and on that plan was the imaging of two large asteroids, Gaspra and Ida. Gaspra is about eight miles in average diameter, and Ida averages nineteen miles across. These were small targets for a spacecraft passing the neighborhood at interplanetary speeds, but Galileo managed to provide the first (and so far only) images of both bodies.

Galileo actually encountered the asteroid belt twice—the first time in a pass through its inner edge, as it continued through its complex series of orbits to gain enough speed to head off to Jupiter. This was when it encountered Gaspra. Because there was some uncertainty about the exact location of the asteroid at the close encounter, Galileo was instructed to take a number of images, fifty-one in all, from which to pick—the asteroid would be in there somewhere. It sounds imprecise, but it worked perfectly.

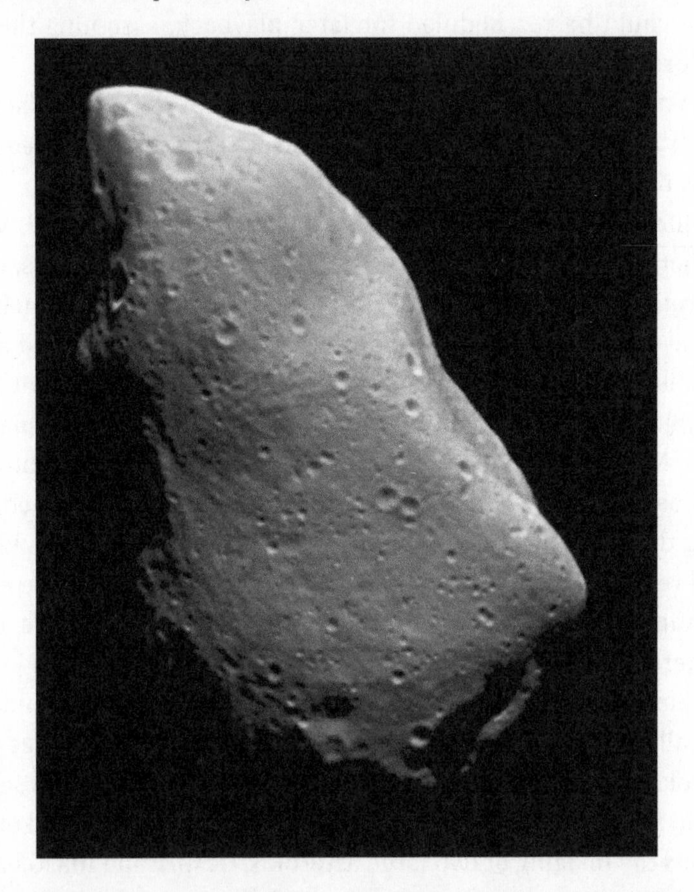

Fig. 29.2. The asteroid Gaspra as seen by Galileo during its flyby. This was captured using a wide-spread collection of fifty-one images. Not bad work for a spacecraft using only its low-gain antenna. Image from NASA.

Then, after passing Earth one more time for an energy boost, Galileo swung out toward the outer solar system for the last time, crossing the asteroid belt and imaging Ida during that quick shot. And when I say quick, I mean just shy of 30,000 miles per hour—which is fast.

The asteroids were photographed with sufficient resolution to make out craters and boulders on their surfaces—even though the gravity of such small bodies is slight, with nothing else nearby to disturb them, rocks and gravel have settled on their small surfaces (Gaspra's surface area is about half that of Hong Kong's). If you were standing there, and had a good arm, you could toss one of those rocks into escape velocity overhand.

Both asteroids were found to be composed of material similar to the most common meteorites found on Earth—a finding that surprised few. One thing that did surprise the science team, though, was the discovery of a tiny moon orbiting Ida, which they named Dactyl (too bad Ida wasn't named Ptera, right?), the first such "moonlet" to be found orbiting an asteroid. More have since been discovered around other asteroids.

And now, the probe headed off to Jupiter. But there was one more major event on the menu as Galileo made its way to the gas giant. About eighteen months before Galileo's arrival, comet Shoemaker-Levy 9 was set to slam into the planet. This was unprecedented, and it offered Galileo a front-row seat to a major interaction between a large celestial object and a planet—something we hope will never happen to Earth. While Shoemaker-Levy 9 was not especially large by cosmic standards, at about two miles in diameter, if such a comet struck Earth the damage would be tremendous—remember that the asteroid that ended the dinosaurs is estimated to be only about triple that size; a two-mile comet would still wreak havoc with Earth's weather systems, not to mention digging a huge hole somewhere (or vaporizing a major volume of seawater, another unpleasant outcome).

Shoemaker-Levy 9 had only been discovered by Earth-bound telescopes in 1993, and was strange even by cometary standards. While it had been a solar system wanderer for a long time—hundreds of millions, if not billions, of years, it followed a bizarre orbit, having been captured by Jupiter's gravitational field just a few decades earlier. But the topper was that the nucleus, or core, of the comet was not a single mass, but was rather a series of separate clumps of ice and gravel, twenty-one of them in all. The comet did not start out that way, but it had been torn apart by Jupiter's intense gravitational field during a close orbital pass in 1992. The once mile-wide cometary nucleus was now a pearl necklace of smaller chunks, destined to burn up in Jupiter's atmosphere at about the same time as Galileo was nearing the planet.

Despite still being 150 million miles from Jupiter when the comet hit, Galileo would have the equivalent of a $10,000 ticket at a Lakers game—a front-row seat. Starting on July 16, 1994, pieces of the twenty-one-chunk-long string of comet junk slammed into Jupiter. And when I say "slammed," I mean it—these iceballs were traveling an estimated 137,000 miles per hour. Mission planners had burned a lot of overtime planning for this, and they were prepared to attempt a number of different methods to image the unprecedented event. Given that the probe was never designed with this specifically in mind, and that the radio dish was crippled, the scientists figured they would be lucky to get much of anything. What a surprise was in store.

With Jupiter's fast rotational period of just under ten hours—the planet spins so quickly that it is flattened at the poles and thicker through its equator—the twenty-one impacts marched across the surface like a strafing run, looking like a series of black wounds puckered across the cloud tops. As the chunks of ice burrowed into the atmosphere, the resulting bubbles of super-hot gas erupted, caused by the intense release of energy, swam back up to the surface. Some chunks made little impact, while others were remarkable—one of these bubbles was estimated to be about two thousand feet across and created a smear across the clouds that was larger in diameter than Earth. The burst that formed it was estimated to have released energies equivalent to many times Earth's global nuclear weapons stockpile.[6]

Fig. 29.3. Galileo image of the impact of a chunk of comet Shoemaker-Levy 9 on July 22, 1994. Image from NASA.

And Galileo was nearby—comparatively speaking—snapping away. Telescopes on Earth were looking toward the planet as well, including the Hubble Space Telescope. Among other discoveries were sulfur and ammonia-bearing compounds stirred up from lower-level cloud layers by the impacts. Perhaps the most lasting effect, however, was not the scars across the planet's surface—these faded over time—but the renewed interest in finding ways to identify and divert similar threats away from Earth. Protecting our planet from a similar, civilization-ending fate is an effort long in the making, but one that has gained traction in the decades since the event.

Galileo sailed on, preparing its assault on the largest planet in the solar system.

Before we fire up those rocket engines to slow into orbit, though, let's look a bit closer at the mammoth planetary probe. Galileo was built by JPL, one of their larger spacecraft to date at 5,650 pounds, and stood over twenty feet tall. It carried two nuclear power supplies, which should be familiar to you by now, using plutonium's radioactive decay heat energy and thermocouples to generate electricity. Galileo had a primary mission of two years, during which it would make an ever-changing series of large orbits through the Jovian system, targeting different aspects of the planet and its moons for observation and inspection.

I've discussed computer processors for other space missions, and Galileo's deserve special mention. Knowing that the Jovian system was saturated in radiation, and that Galileo would be loitering there for years, the JPL engineers had not one, not two, but six processor chips fabricated—on sapphire wafers no less, ideal for dealing with high levels of radiation and static electricity. The chip was of about the same processing power as the Apple II computer of the 1980s. It had a paltry 16k of RAM.

Once again John Casani was called upon to help craft a solution to the ongoing issue of radiation protection:

> What we found out on Voyager was that the radiation intensity in Jupiter was going to be greater than the capability of the parts to survive, so we instituted a shielding program to put mass between the radiation and delicate parts. The spacecraft structure provides some mass, but after a lot of analysis we realized that you just can't completely interrupt the fields internally. In the end we had to resort to putting pieces of equipment in strategic locations where they would have some protection. We also had some local shielding . . . little titanium hats that we put over certain components.[7]

The goal was to surround the computers with as much equipment as possible—equipment that would shield the more delicate computer chips from much of the ambient radiation.

Because the spacecraft would be the first to stop and orbit an outer solar system planet, it had a main rocket engine to augment the twelve maneuvering thrusters arrayed about its structure. Without this, Galileo would have sailed merrily past the planet, but in December 1995 the engine was fired to slow it enough to be captured by the giant planet's gravitational field, and Galileo's complex series of orbits through the Jovian system began, each taking about two months.

But there was more to Galileo than the main spacecraft—it also carried a separate probe that would detach before it reached the planet and eventually descend into Jupiter's atmosphere. This atmospheric descent probe weighed about 750 pounds and was a four-foot-wide cone with a dense heat shield that accounted for almost half that mass. Inside the probe's shell, which was designed roughly along the same lines as the protective reentry covering of a nuclear warhead, were a variety of sensors and instrumentation. As you would expect, temperature, pressure, and radiation were to be measured as it plunged deep into the clouds of Jupiter, along with electrical fields and the types and abundance of energetic particles.

The descent probe, which was rather uncreatively named the Galileo Probe, detached from the main spacecraft about five months before its mothership would encounter Jupiter, headed on a different trajectory that would send it into the giant planet itself. As the descent probe neared the planet, it discovered a new radiation belt surrounding Jupiter, close to the cloud tops.

On December 7, 1995—the same day the main spacecraft reached Jupiter—the probe slipped into the planet's gaseous envelope at a speed of about 108,000 miles per hour, or about Mach 145 (for those of you prone to wonder about such things). The shield experienced heating of up to 28,000 degrees Fahrenheit. Peak deceleration forces were about 228 g's, far heavier than any previous spacecraft had ever encountered. This was one robust little machine.

Once the probe had slowed sufficiently, a parachute was deployed to allow it to drift through the complex vortexes of windy cloud bands surrounding the planet. It operated for about an hour as it fell through the first hundred miles, failing only when atmospheric pressures exceeded twenty-three times Earth's. The probe found that the upper layers of the atmosphere had less helium present than expected, but they were far windier—about 400 miles per hour. Later analysis indicated that the probe may have fallen into a hot spot in the planet's atmosphere. Nonetheless, the measurements were telling.

Torrence Johnson was the project scientist for the mission, a job he described as being "the chairman of the board" for the other scientists on the mission—it was equivalent to Ed Stone's job on Voyager. He noted that what the drop probe found was not quite what they had expected. "Jupiter was generally thought to have been built of material from the solar nebula with solar composition," he said—the solar nebula was the stuff the solar system was originally made from. But a number of elements were measured at much higher rates than expected: "Carbon, nitrogen, and sulfur as well as the noble gases argon, krypton, and xenon are all enhanced over expected solar abundances by a factor of about three. This suggests significant enrichment of the solar nebular gas by accretion of cold icy planetesimals [small, primordial, planet-like clumps of detritus] early in Jupiter's history and represents a key constraint on the conditions and timing of gas giant formation in the solar system."[8] In terms that don't cause the rest of us to hear a dial tone in our heads, the elemental differences found in the drop probe's brief life seem to indicate that the gas giants, and the early solar system, may have formed in different ways than previously thought—there were more icy proto-planets hanging around, waiting to be swept up like juvenile delinquents in a late-night police raid by megacop Jupiter, of which they eventually became a part.

On that same day, the main Galileo spacecraft swung by the moon Io and fired its main engine to place it into its first large orbit through the system, even while collecting and preparing to retransmit data from the smaller drop probe.

With everything deployed successfully (except for that pesky main antenna), Galileo went about its business. Magnetometers mounted on a thirty-six-foot boom began measuring the environment, held clear of interference from the main body of the spacecraft. Instruments to measure charged particles and their waveforms were also in place. Other devices measured the heavy radiation surrounding Jupiter and the amount of UV present.

Another section of the spacecraft, called a scan platform, which could move independently of the main chassis, contained cameras for visual imaging, spectrometers tuned to both UV and infrared wavelengths, and photopolarimeters to measure illumination in this dim part of the solar system. The cameras were particularly noteworthy, as they were capable of providing images far superior—up to one thousand times better, by some measures—than the Voyager cameras. A lot can happen in a decade of design and improvement. This was one of the first modern CCD (charge-coupled device) cameras flown, providing images of 800 x 800 pixels, which was considered high definition at the time.

An aside here—once again, as with the early Mariners, NASA had considered a mission to a planet, and a visually stunning one at that, without a camera. "Galileo was originally conceived to not have a camera," Casani says, "but the remote sensing community went bonkers."[9] Bruce Murray, with whom Casani had worked during the Mariner missions, was now JPL's director, and he favored having cameras on Galileo—but Casani had to make sure they would fit onto the spacecraft, both in terms of placement and mass: "We would get together with the science teams and figure out what they needed, how much performance, how much mass they would have, all that kind of stuff. My job was to moderate this. I had to make the spacecraft work, and also had to make sure that all the science instruments got what they needed—but not *more* than they needed, and not to the point that the spacecraft wouldn't work. I was trying to balance everything."[10]

For the next eight years, Galileo performed thirty-five orbits, weaving through the busy Jovian system. These elongated orbits not only allowed the spacecraft to do close inspections of a number of moons and belts surrounding Jupiter, but they also minimized its time in the most radioactive zones of the planet's orbit.

There were many noteworthy discoveries in Galileo's long list of accomplishments. One exciting episode occurred when the probe swung past Ganymede early in the mission. While the moon had been imaged by Voyager during its blistering pass through the system, Galileo would enjoy a far more detailed view. A region called Galileo Regio was of particular interest, as the Voyager images had captured a series of smooth grooves in one area—the entire region was puzzling, but this spot was fascinating. Said Torrance Johnson, "High-resolution images of nearby Uruk Sulcus, more than ten times better than Voyager pictures, quickly resolved the origin of the bright, apparently smooth lanes that Voyager had seen. Galileo's images revealed that, seen close up, they were not smooth frozen flows of icy 'lava,' as had been suggested after Voyager, but rather a tectonically reworked region where dark material had slumped off the steep slopes of a set of faulted and fractured ridges."[11] In other words, Ganymede appeared to be geologically active. This finding was supported by the detection of a dense magnetic field there, indicative of a molten, iron-rich core—the first moon in our solar system found to have one.

Fig. 29.4. Fantastic, close-in images of Jupiter would provide planetary scientists material to study for decades. Shown here are four different images of Jupiter's Great Red Spot, a hundreds-of-years-old cyclonic storm that is a main feature of the planet, shot at different wavelengths of light. Image from NASA.

Io was also a target of intensive investigation due to the astounding images sent back by Voyager. However, as the closest moon to Jupiter, Io was a risky target—the radiation that close to the planet was potentially crippling. It was decided that it was worth the risk, however, and with images coming in at many times the resolution of Voyager (despite Galileo's crippled antenna) the photos showed that the volcanic eruptions seen by Voyager were continuing unabated. This aroused the mission scientists' interest (frankly, *everything* did, but Io is particularly spectacular). They wanted a closer look—but what about that damned radiation?

Over time, it became clear that Galileo was resisting radiation damage better than the engineers had thought likely, and, as mission milestones were met, multiple closer passes to Io were planned and executed with impressive results. There were eruptions happening at that very time, said Johnson. "Estimates from images and spectra of the temperatures associated with eruptions confirmed that the majority of Io's volcanic flows are molten rock at very high temperatures, similar to basaltic lavas on Earth."[12]

But the best was yet to come. Europa was possibly the most intriguing moon, despite the active conditions on Io and the discovery of a magnetic field on Ganymede. The greatly improved imagery from Galileo showed a very young and chaotic surface on Europa. The lack of cratering on the surface indicated how young that surface was, if you can call 50–100 million years young. But in a solar system that's 4.5 billion years old, 50 million years is a trifle. When you turn fifty, visit your Uncle Morrie at a meeting of the Elks Club, and you'll know what I mean—those gents won't seem quite so elderly as they did when you were thirty. Age is a relative thing.

But young did not mean undamaged. Despite the lack of cratering, there was still plenty of dramatic landscape to be seen on Europa. There were wide regions that showed a history of violence—broken and tilted blocks of ice, looking much like Earth's arctic icefields, except frozen like steel in the deep cold of an outer solar system moon. "Gravity data taken during close flybys also indicated a low density . . . layer about 100 km deep overlying a denser rock and iron interior. These observations were all consistent with models showing that a liquid water ocean, with a volume twice that of all Earth's oceans, could be maintained under an icy crust by tidal dissipation," Johnson said.[13] Let me repeat: this little moon of Jupiter may contain as much water as all Earth's oceans, times two. Wow.

But what causes all this violence and mayhem on Europa's surface?

Yes, you guessed it—the evil, clutching hand of Jupiter. The planet's intense gravitational field squeezes and flexes the moon, according to Robert Pappalardo, who worked on the imaging team for Galileo fresh out of grad school. "As it orbits around Jupiter, Europa 'breathes' in and out, flexing in its slightly elliptical 3.55-day orbit about Jupiter. Europa is tidally stretched more as it gets closer to Jupiter and contracts as it moves further away. If there is indeed a decoupling ocean between the outer icy shell and the deeper interior, then the radial tide should allow the surface to rise and fall by a total of 30 meters with each orbit."[14] That's almost 100 feet—a huge amount of vertical displacement for a sold surface, even one with some elasticity, such as ice. You can see why Europa is warm, with all that squeezing and squishing, and why its surface looks like it went face-first through a plate-glass window in its youth.

Europa turned out to be a wet, possibly warm ocean world with a frozen crust atop the water—and potentially a perfect place for life to arise. Ganymede and Callisto may have subsurface oceans as well, and more such watery moons would be found around Saturn.

Like Mars after the Viking missions, the solar system had become a much livelier place since the first tentative observations of the early space age. What had been assumed to be dead, lifeless, and mostly inert worlds were taking on a new life—at least some of them—as ice-covered moons with warm oceans below.

As the years wore on, Galileo started to wear out. The radiation around Jupiter was taking a toll, and instruments were starting to malfunction, though there were no total failures. Cameras would intermittently go white, a quartz crystal used for a radio frequency baseline reference shifted frequency when the spacecraft got too close to Jupiter, and the computer began resetting itself. As noted, electronics and radiation go together like a marshmallow in a blast furnace—not a happy pairing.

Galileo's daredevil swings past Europa popped back into the news a few years ago. There were eight passes of the fascinating moon by the probe in total, and two of them were at low altitudes of less than 250 miles. That's some pretty good shooting when you're controlling a spacecraft from an average of 390 million miles. At any rate, the readings the probe recorded as it passed nearest the surface were a bit weird—but since the data coming back from Galileo were not always crystal clear, these measurements were shelved for later analysis—it was thought they might simply have been equipment malfunctioning from radiation damage. Then the data were reexamined just a few years ago.

New analysis shows that the close dives of Europa, which took place in 1997, corresponded with some "hot spots" that were later detected. The anomalous readings from Galileo—a shift in the magnetic field and a rise in plasma density—were not instrument malfunctions or garbled data. The spacecraft had apparently flown right through one of the Jovian moon's recurring plumes. To make the case more compelling, the position of the spacecraft during that part of the flyby correlated nicely with a Hubble Space Telescope image of a plume occurrence. All this is compelling enough to add to the already many good reasons for a Europa orbiter mission, called Europa Clipper, which is currently in the planning stages at NASA.

After fourteen years of sterling service around Jupiter, Galileo was showing its age. Fuel was low, and systems were becoming increasingly unreliable. It was time to put the old Jupiter orbiter out to pasture. There aren't a lot of choices about how to end a mission; if you simply do nothing, the spacecraft will eventually fail or run out of fuel and lose alignment with Earth, in which case it will begin to tumble along its endless orbit until it either smacks into something or is drawn in by Jupiter's gravitational field. Either could take a long time, but the prospect of Galileo inadvertently slamming into one of Jupiter's moons was not a thrilling one—the craft was not sterilized to the kinds of standards that NASA applies to missions that might come into contact with water-bearing worlds, and nobody wanted an old, defunct space probe that was covered in Earthly microbes polluting a distant moon that might have some form of indigenous life.

The other option? Deliberately destroy your multi-billion-dollar orbiter. Check this box and sign on the dotted line, please.

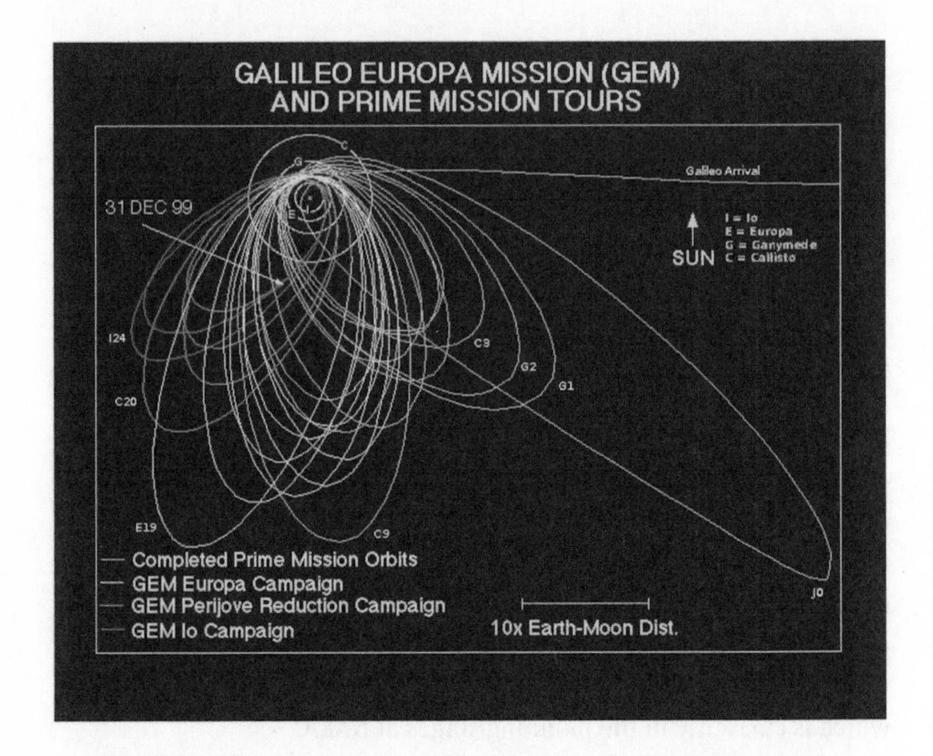

Fig. 29.5. Galileo's orbits through the Jovian system. Navigating the complex region around Jupiter was also complicated by the diminished bandwidth of the small low-gain antenna, but mission planners figured it out, and the results were stunning. Image from NASA.

After eight years at Jupiter, and in its thirty-fifth orbit, Galileo flew past the small moon Amalthea, swung out on a larger orbital leg than usual—about 16 million miles from Jupiter—then, like a bowling ball shot into the sky, returned in a downward plunge, this time aimed squarely at the giant planet. On September 21, 2003, after traveling over 2.5 billion miles, Galileo was incinerated by Jupiter's atmosphere, entering just south of its equator at a speed of about 108,000 miles per hour.

Over a hundred of the Galileo mission team's participants joined the press and various JPL and NASA staffers for those final moments.

The end of a mission is always a mixed bag—the pride of accomplishment is blended with a wistfulness that the mission cannot continue. Of course, there are a select few in NASA management that quietly prize such moments, since a mission termination is a cost item can then be removed from NASA's balance sheet—ongoing mission management is not cheap. But most of those involved are more philosophical.

As Torrence Johnson said at the mission's concluding press conference in 2003, "We haven't lost a spacecraft, we've gained a steppingstone into the future of space exploration."[15] Johnson also recalled a favorite memory of the Galileo project at another press event, this time with a smile: "Following the end of Galileo's prime mission, a conference was held in Padua, Italy to commemorate the three Galileos: the man, the spacecraft, [and] the telescope. At the close of the meeting, Pope John Paul II received the participants and guests in Rome on January 11, 1997. His Holiness's succinct reaction on viewing one of Galileo's pictures of Europa is hard to better as a summary of Galileo's mission: 'Wow!'"[16]

That kind of brief but enthusiastic approval from the top executive in a completely different branch of study meant a lot.

Chapter 30

FLASH FORWARD: INFLATABLE ANTENNAS

After the Galileo mission, and with more spacecraft under construction that would require ever-higher bandwidths, NASA returned to solid, one-piece radio dishes for spacecraft like the Cassini probe to Saturn. Nobody wanted a repeat of Galileo's failed antenna deployment. But, as always, the maximum diameter of the radio antenna was dictated by the width of the rocket fairing that launched it, putting constraints on how much bandwidth the transmitter would have. Enhanced compression methods would help, as would improved ground hardware, but there should be a better way to devise larger antennas for planetary spacecraft in the future. As it turns out, there is—NASA researchers at JPL have been investigating this for years, and they are on the verge of a number of breakthrough technologies that will provide for much larger antennas packed in smaller spaces for launch. Some unfold, some unfurl, while still others . . . inflate.

The latter is a simple idea: coat one-half of the inside of an inflatable plastic ball with a reflective surface, which reflects radio signals that enter through the other, untreated half of the sphere back to a collector, allowing the ball to act as an antenna. Ideally, the inflated structure would be a parabola, but such shapes have proved difficult to deploy reliably. After some experiments with attempted parabolic inflatables that were less than satisfactory, the developers of these beach ball-like antennas have used computer modeling to mimic that shape using a signal interpretation software.

Such designs can be utilized as both traditional spacecraft

antennas and for space-based radio astronomy applications. Inflatable antennas have been tested at about three feet in diameter, so that even a CubeSat could carry one antenna several times its own width. The design is also scalable, suggesting that much larger antennas can be successful—for example, a three-foot wide spacecraft might deploy an antenna several times its own diameter.

Deployment is simple—once the spacecraft is in orbit, a highly compressed plastic sphere is released and is inflated until taut. Researchers are also looking at ways to rigidize the sphere once it is pressurized, by using coatings that harden when exposed to ultraviolet light, at which point it would no longer be prone to changes in shape when punctured by microscopic debris or due to temperature fluctuations, which are extreme in space.[1]

Besides the antenna's obvious utility for communications, larger antenna designs would greatly increase the sensitivity of orbiting radio telescopes. Future missions might include mapping the abundance of water in space, which is of interest due to its importance with regard to the development of life. In addition, astrophysicists want to follow the "water trail" from interstellar clouds, to solar system objects such as comets, the Earth, and other worlds. But so much water is present in our atmosphere that ground-based observations are impossible; these must be conducted from space. Also of interest would be a CubeSat utilizing inflatable antenna technology to make an all-sky survey of the interstellar medium in the Milky Way. A satellite that included a roughly three-foot antenna could measure the motions of thousands of interstellar clouds in and around the Milky Way and lead to a better understanding of our galaxy and its surroundings.

Inflatable antennas are simple to deploy—just push them out of the spacecraft and inflate. No moving parts, such as ribs and vanes, are needed. Using this technology for spaceflight operations, especially small CubeSats that are challenged for antenna space, is a natural progression.

Sometime in the not-too-distant future, the familiar dish design of radio antennas may be replaced by something that appears to be an oversized monochrome beach ball, so be prepared—no kicking allowed.

CHAPTER 31

THE CENTER OF THE UNIVERSE: PART 4

By all accounts, Curiosity had landed on Mars and was ready to prepare for its mission of exploration. At least that's what the clocks said—but not what the readouts relayed. With the delay between occurrences on the red planet and the reception of the radio signal here on Earth, nobody at JPL would know for fourteen more minutes. It was like a replay of the Viking landings in 1976, or the Pathfinder landing in 1997, or that of the MER rovers in 2004 . . . wait and see.

In the SFOF, NASA folk from the top down were quietly biding their time. Charles Bolden, the NASA administrator was there, as was Charles Elachi, the director of JPL, and others. There was little to do now other than wait. A few in attendance absentmindedly chewed fingernails or wrung their hands as they looked at control screens of the monitors overhead, counting the minutes until confirmation.

In Von Karman Auditorium, the chatter was subdued. The younger folks, many of whom were new to space journalism, reacted in ways ranging from jumping out of their skins to checking email and tweets on their phones. The more veteran of the media crowd, having been through a number of such exercises, sat with arms crossed, staring at the big screen at the front of the room, or swapping war stories with colleagues in low tones. But everyone had an invisible clock counting backward toward zero inside their heads. The moment of truth had almost arrived.

Across Pasadena, at the Caltech campus, the auditoriums were filled to overflowing, and huge inflatable video screens had been set up on the grass outside to host thousands of people who had come out on a pleasant evening to witness history in the making. There were picnics and small children running around, some mimicking the sky crane descent, as previously modeled onscreen by one of Curiosity's engineers. It was about ten p.m., but felt much earlier—nobody gave a moment's thought to heading home before the first images came in. The wider US audience, watching on network news or on streaming video, was conservatively estimated at well over three million.

And so it was across the country, and in many others, where people gathered in public places to watch large-screen reporting of the landing as it unfolded. Onscreen, seated with six others in a row of blue-shirted mission controllers, was the now-iconic image of "Mohawk Guy," Bobak Ferdowsi, who made JPL and NASA cool again. With his signature Mohawk haircut, complete with multicolored stars shaved into the thin hair along the sides, the handsome young engineer cut quite a remarkable and, to many, unexpected image among the other, more staid members of his team. He would receive a few thousand admiring tweets that evening; in fact, he was receiving them at that very moment—including a few marriage proposals. When I spoke to him afterward, he noted that had not yet responded to the latter.

As Ferdowsi said a bit later, "Of course, we had had many dress rehearsals before that to make sure we knew what we were doing. So we just fell into the routine that we had used for most of cruise [Curiosity's trip from Earth to Mars]—there are procedures in front of you, polls are taken and you respond to the polls, see what the values are for the part of the software that you are looking at. In that way it is sort of calming, because it is a routine that you have been doing this whole time." But of the landing, he said, "At this point it's all happening fourteen minutes ahead of what we are seeing, so, for instance, we have just jettisoned the cruise stage, etcetera, so things are actually happening, *right now*."[1]

Al Chen was the "voice of Curiosity," manning the console that provided the moment-by-moment verbal coverage of the landing. Though he looked calm enough, he was nervous as hell. All that was being relayed back from Curiosity during landing was a series of telemetry tones, called a "heartbeat," that told only the most critical parameters of the probe's status. "These 256 tones that Curiosity would send back tell us what was going on. That's all we had for about 14 or 15 minutes, from cruise stage separation until we got into the atmosphere. After that we started picking up data from Mars Odyssey."[2] NASA's Mars Odyssey orbiter was serving as a relay station for the landing. For the first few minutes, everything looked good. Then, a problem—though you would not have noticed it by watching Chen on the video screen. All he showed outwardly was a furrowed brow.

Fig. 31.1. Bobak Ferdowsi, a JPL flight controller for the Curiosity landing, became an internet sensation overnight. He went on to participate in a lavish PR campaign for NASA/JPL and was a media darling for the next few years. Image from NASA/JPL-Caltech.

"Shortly after we hit the atmosphere we got a tone, called 'Data Catastrophic.' It's usually bad when you get these tones coming back that say 'catastrophic,'" he said wryly. "We had intended that tone to tell us that we were about to lose the vehicle. . . . I almost had a heart attack when I saw that. If it was true, it meant either that the vehicle was tumbling or it was not heat shield forward," meaning that it would burn up as it plummeted deeper into the Martian atmosphere.[3]

The landing appeared to be in jeopardy. Then—

"I thought about it for a second. . . . I thought, if we are about to lose the vehicle, I'd better not say anything dumb, going back to guideline one. I thought, this should happen right when we hit the atmosphere, there should not be much error at that point. Do I really believe this? Is there more going on there than what I was thinking about? . . . We were about to hit peak deceleration and peak heating, so it will become obvious if we are about to lose the vehicle in a minute or two . . . so I sat on it. Luckily it turns out that the instrument was working fine, we just had a calibration error. If I was not already sweating then, I sure was after that."

Von Karman grew very, very quiet as we all stared at the screen, willing the spacecraft to send us the proper message. Chen continued his intermittent narrative from mission control.

"We're now in powered flight . . . down to one kilometer and descending . . ." he said in hushed, almost flat tones.

Curiosity had shed its heat shield and the sky crane rocket pack was firing . . . or, more precisely, it had been firing fourteen or so minutes earlier.

"Five hundred meters in altitude . . . standing by for sky crane." This announcement signaled the beginning of the rover's rappelling down from the hovering rocket pack. Then, after a few moments, "Sky crane has started, descending at about point-seven-five meters per second as expected." A bit of clapping from mission control, tense stares from the rest of us. We didn't build it, and we weren't flying it, so the meaning of the moment was not as profound.

As this point on the secondary video feed, we could see the faces of the controllers, with Adam Steltzner, the lead engineer on the EDL phase of the mission, pacing behind the back of the front row of controllers like a caged tiger. He would stop for a moment, looking intently over someone's shoulder at a computer screen, then continue his patrol.

Then, Chen said, "Touchdown confirmed—we're safe on Mars!" And the room went wild, as did the feed from mission control. They had done it, and all was well—probably.

Fig. 31.2. Rob Manning was the chief engineer behind Curiosity. His team sweated out the final months of preparation before the landing, testing everything exhaustively. Image from NASA.

According to Curiosity's chief engineer, Rob Manning, there was a final step before everyone could really, truly take that deep breath they'd been hoping for over the past number of years: "When the wheels made contact, the rocket pack continued coming straight down, slowly, so it never really hovered then. When the rover sensed ground contact, the computer realized that the rockets were using less power to stay up in the air. The computer then said, 'I must be on the ground.' So it sent the command up to the rocket pack to say 'stop and hover,' so *now* the descent stage begins to hover. It slowed to a stop and finally the computer cut the cables. Before it cut the last electrical cable, it said, 'Okay we're done, go ahead and fly away!' and in one second the descent stage replied 'Right—copy that.'"[4] This all took place in a few seconds, but these final steps were critical.

So, while most were celebrating, a few of the engineers were continuing to watch the readouts to make sure that the rocket stage had gotten the message and separated completely. More than one engineer—including Manning—had endured nightmares of just one of the tethers failing to separate, and having the rocket pack fire to gain altitude and get away from the rover, but dragging Curiosity along with it...

A few second later, separation was confirmed, and the rocket pack had flown a few miles off to crash in the distance. Curiosity was safe and sound, sitting on the floor of Gale Crater. Now *everyone* could stand up and cheer.

Chapter 32

SATURN AHOY

On September 5, 2017, the faithful made the pilgrimage back up to JPL for a seminal event. After nearly twenty years in space, the Cassini Saturn mission was coming to an end. NASA had decided, as they had with Galileo before it, that with a moon system full of interesting worlds to explore—a number of which could harbor life in their subsurface oceans—the risk of the probe drifting uncontrolled through the system was a concern. Cassini could tumble out of control once its fuel was exhausted and accidentally slam into one of those watery worlds, contaminating it and possibly ruining the prospects for future discoveries. So Cassini was sent on a final, one-way trip into Saturn's atmosphere. There, any germs on the probe, and its plutonium fuel source, would do no damage as the spacecraft burned up in the upper atmosphere.

The weather was moderate in Pasadena—summer stretches from June through October here, so temperatures below ninety were a rarity—and when I arrived at JPL just before midnight the mall was as pleasant as could be. I had just finished writing a technical book for JPL a few months earlier, and my badge would expire soon anyway, so I thought I might as well get one final use out of it for this mission-ending event.

As usual, the press had jammed the guest parking lot, so I parked in the western reaches and wandered over to Von Karman Auditorium. Even at midnight, the place was humming—many folks had shown up for the final act of this flagship mission.

As we settled in for the five-hour wait, the press took their usual spots with the TV crews near the back, the social media set behind them, and the rest of us scattered throughout the auditorium. Space writers recognized one another, and conversations about new developments in spaceflight focused largely on rumors and speculation about the nine-month-old Trump administration's priorities for NASA and a new administrator for the agency. We thought we'd have one very soon—an incoming president typically announces the new administrator within six months of taking office; Trump was taking far longer. As it turned out, we'd have far longer to wait, and many people's guesses about the new chief of the agency were way off base. (Oklahoma representative Jim Bridenstine ultimately got the job.)

On center stage tonight was the final plunge of Cassini into Saturn's roiling cloud tops, which would occur just before dawn. For the past several months, Cassini had been making ever more daring dives between the planet and its innermost ring, getting a bit closer to the planet on each pass. The gap between this inner ring and the planet was just 1,500 miles across, so, while there was ample room for error, this was a high-radiation region that had not been previously explored.

The views had been spectacular, as had imagery throughout the course of the landmark mission. No other planet offers the kind of spectacular visuals that Saturn does, and the amazing landscapes of its moons were an added bonus. It was an incredible mission.

Cassini's final orbits had been short, about seven days each, with each successive pass bringing it closer to the planet's upper atmospheric layers. By 12:30 p.m. California time, Cassini sailed past the farthest point of its orbit, well beyond the outermost rings, and was now making a true final dive toward the planet. At 77,000 miles per hour, Cassini wouldn't last long once it hit the upper atmosphere. For the last fourteen and a half hours of its life, it had kept its radio transmitter aimed at the Earth, and would continue to do so until the end, transmitting data in real time—each moment of science in this grand finale was too precious to lose.

Cassini, NASA's first Saturn orbiter, had launched two decades earlier, on October 15, 1997. There would be no more messing about with the space shuttle as they had with Galileo; Cassini went aloft on a Titan IV rocket, an evolution of the long-serving Titan ICBM from the 1960s. The larger, evolved Titan was required for the heavy probe, which when fully fueled and ready to go weighed 12,600 pounds; when dry, the combined orbiter and drop probe would be a bit less than half that.

The prelude to Cassini's launch was not without its challenges, however. The spacecraft was being built during Dan Goldin's tenure as the chief of NASA—the same man who had spearheaded the "faster-better-cheaper" mantra at the agency and who had presided over a number of mission losses in the 1990s. First, the Mars Observer mission fell silent in 1992 before reaching the red planet.[1] Then the Mars Climate Orbiter failed in 1999, and the Mars Polar Lander crashed in the same year. The latter two mission failures are thought to be due primarily to cost-cutting measures. Despite this less-than-stellar record, Goldin was still not a fan of big "flagship" missions, and he pushed back against the ongoing development of the massive Cassini orbiter.

Julie Webster, the chief engineer on the project, summarized his views succinctly: "Goldin was calling us 'Battlestar Galactica.' We were huge in an era when they were trying to go to better, faster, cheaper. He didn't like having all the eggs in one basket like that."[2]

Goldin grudgingly went along with Cassini, but the team still had to sell it to Congress. This would involve, among other things, demonstrating a willingness to compromise on costs. John Casani, again appointed as a project manager during Cassini's trying years (there were a number of them over the course of the mission), put it this way to his team: "We'll show that we can bleed," and with that, the scan platform was removed as a cost-cutting measure.[3] Scan platforms are motorized mounts that move independently of the body of the spacecraft, allowing specific instruments to be pointed in directions that the spacecraft is not facing. This saves fuel, for without it, the entire probe needs to be reoriented for these instruments to do their jobs. But it reduced the budget by about ten percent, and that's what mattered, so *poof* went the scan platform.

Linda Spilker, who joined the Cassini project in 1990 and later became the project scientist through the end of the mission, elaborated: "One thing we did that I thought was very smart is we set up a trading board and allowed the instruments to trade [each other] for mass, power, data rate and money."[4] There is always a scramble in the planetary science community—both inside and outside of NASA—to put more numerous and increasingly capable scientific instruments on robotic space missions, especially when an opportunity as rare as this presents itself—Cassini would be the first spacecraft able to loiter in the Saturnian system (only the Pioneer and Voyager missions had visited previously, and both via flybys), and likely the last to travel to Saturn for decades to come. This was their big chance! But in the US space program, budgets rule the day, and this one had to be trimmed. So the trading board was a system by which to swap instruments and their capabilities for mass and cost. It was a brutal and bruising process, but necessary. As Earl Maize, a subsequent program manager who stayed with Cassini until the end, said, "In all missions there is a fixed amount of power, a fixed amount of mass, and a fixed amount of data bandwidth. So the instruments were allowed to trade various components. For example, I need 2 watts, but I don't need 2 kilograms—does anyone have an extra 2 watts for 2 kilograms?"[5] It's a simplification of a complex process, but you get the idea. Cassini was the first time such a system of trades was formalized in this manner, but it's been used many times since.

Cassini came together slowly, and NASA was looking for ways to reduce costs from the beginning. An early savings resulted from collaboration with Europe. In 1982, when NASA was just starting to look formally at a Saturn orbiter, the European Space Agency was soliciting general mission proposals from European scientists. A young planetary scientist named Wing-Huen Ip, who worked at the Max Planck Institute in Germany, brashly proposed a Saturn orbiter with an atmospheric probe to plunge into the clouds of either Saturn or Titan. He knew it was a long shot, and he called a colleague in France to discuss possible collaboration. Others got involved, and the project grew legs at the European Space Agency.

Fig. 32.1. Linda Spilker, project scientist, speaks at a press conference at the conclusion of Cassini's mission. To left is Earl Maize, the program manager. Image from NASA.

Meanwhile, in the United States, NASA was conducting workshops to define goals for solar system exploration. Toby Owen, a scientist at the University of Hawaii, who was also working on early Saturn orbiter proposals, was chairing the Outer Planets Group. While the group was not initially approved to include foreign partners, Owen knew some of the people working on similar Saturn proposals in Europe, and by 1988 a partnership was formed to blend an American-built orbiter with a European atmospheric probe called Huygens. The probe would be targeted to Titan, Saturn's largest moon. This plan would (it was hoped) help to save more money by spreading costs between the two space agencies, and it had political overtones as well—the Soviet Union was working more closely with ESA than ever before, and NASA did not want to lose a seat at the table.

There was one more component to this evolving plan, which involved blending yet another set of mission goals: build a twin to

Cassini, as had been done with the Mariner and Viking spacecraft, to explore a comet. This mission was called CRAF (for Comet Rendezvous Asteroid Flyby). The overall mission plan was now called CRAF-Cassini, and by 1989 it was working its way through channels that would ultimately require the dreaded congressional approval.

But by the time the dollars were totted up, the cost of the combined missions was too high, and the CRAF component was canceled in 1992. It would be interesting to one day add up the total costs of the studies and proposals that NASA has *not* flown, but that's a topic for another time.

It was at this point that Casani's edict to make the mission "bleed" was enacted, and with fervor. Instruments were "de-scoped," NASA-ese for making them smaller and less expensive—and sometimes less capable. But planetary scientists and engineers are a clever lot, and given time will often find ways to come close to, and sometimes even exceed, their original goals with less lavish levels of technology.

By the time Cassini launched, twenty-eight countries were involved in various aspects of the mission, with the main Cassini spacecraft built at JPL and the Huygens drop probe assembled in France. The Italian Space Agency provided the radio dish (a solid dish this time—there would be no repeat of the Galileo debacle), a radar, and other instruments.

The primary objectives of the Cassini mission included:

- Understanding the structure and dynamics of Saturn's mysterious rings (only hints had come from Voyager).
- Imaging and investigating the moons of Saturn and understanding their geological processes.
- Investigating the dark material seen on one side of Saturn's moon Iapetus in Voyager photos.
- Study of the complex behavior of Saturn's banded atmosphere.
- Study of the atmosphere of Titan—both from above (by Cassini) and from within (via Huygens).
- The characterization of Titan's surface, at that time unseen due to the dense blanket of clouds that surround the world.

This is an incomplete list, but it covers the main topics of interest. It was asking a lot of one orbiter and a single drop probe, but in the end Cassini would deliver this and so much more. The budget was a hefty $1.4 billion up until launch, but in light of Galileo's cost right in the ballpark. It's worth noting that $422 million was for the Titan launch vehicle—it was not a cheap ride into space, but it was necessary given Cassini's mass and size. Cassini measured thirteen feet across by twenty-two feet in length, larger than a standard cargo shipping container.

While sending a single, large spacecraft may seem risky and expensive (especially to someone who characterized it as a "Battlestar"), missions like Cassini are somewhat akin to that long-awaited trip a New Yorker wants to make to Mexico with her family—getting there is expensive, but *staying* there less so, which means that if you look at overall economies it makes sense to extend your vacation (and if John Casani is your tour guide, he might suggest skipping the guided tours of the Mexican pyramids and direct you to "Just walk around and look for yourselves . . ." to save money). The cost of the mission over two decades of operations was about $3.6 billion.

Despite the use of the powerful Titan rocket, Cassini still needed some help to head out to Saturn. After launching in 1997, it flew past Venus twice and Earth once, then sped toward the outer solar system. After crossing the asteroid belt, the probe headed past Jupiter, snapping some of the best global images of that planet yet. Though it passed the planet at a distance of over six million miles, the improved cameras aboard provided a stunning view.

Cassini was far more complex than Galileo and built on much of what had been learned on that mission and those of the Voyagers. Once again, a militarized computer processor would be the brains of the spacecraft, a sixteen-bit chip first developed in 1980. Multiple units would be used for redundancy, necessary given the radiation-rich environment in which it would be working. Also like its other deep-space siblings, Cassini would be powered by three plutonium fuel sources to generate electricity.

The science package contained what is surely by now a familiar set of instruments: a spectrometer that operated in both infrared and ultraviolet light, a cosmic dust analyzer, a magnetometer to measure magnetic fields, a mass spectrometer to analyze particles found in the Saturnian neighborhood, charged particle detectors, a synthetic-aperture radar to map the surface of Titan (synthetic-aperture radars use the movement of the radar receiver to synthesize a larger conventional dish), and more. Also included were improved CCD cameras—there was no argument about cameras this time; images would be returned to Earth to make both the scientists and taxpayers happy.

The Huygens drop probe was a 773-pound unit that looked like an oversized (4.5-foot) metal Roomba vacuum cleaner (of course, Roombas had not been invented yet—but the comparison is apt). It was battery driven and would only have a few hours of life on Titan, but what hours those would be. Huygens contained sensors for measuring the properties of Titan's atmosphere, and acceleration sensors to measure wind gusts and changes in direction and velocity as it plunged through Titan's dense, toxic atmosphere. It also measured temperature and charged particles, and even carried a microphone—the first to the outer planets.

The radio had a special oscillator included to allow the radio science team to measure horizontal travel of the probe via Doppler shift, the same thing that makes a police siren descend to a lower tone when it passes you. Huygens carried a gas chromatograph/mass spectrometer to identify key elements in Titan's atmosphere, and other instruments targeted at a limited analysis of the surface on which it landed, whether liquid or solid—nobody was sure what was down there, but both types of surfaces were hypothesized. Finally, it carried cameras to take images of the descent and the landing site. The small probe would transmit data directly to Cassini, which would record and relay the information to Earth.

Cassini arrived in the vicinity of Saturn in June 2004. Within months, the spacecraft had identified some new and unknown moons; seven were discovered during the mission. Like other solar system moons, these were named after figures in Greek mythology of ever-diminishing importance (NASA was discovering a lot of moons in the decades since Voyager). These included Methone, Pallene, Polydeuces, and Anthe. Two more were found in the rings that surround the planet, Daphnis and Aegaeon. A later discovery in 2009 netted another moon, which is currently called S/2009 S1—it needs a better name, but at only about a thousand feet across, it can wait.

Early in its visit to Saturn's neighborhood, Cassini made its only flyby of the moon Phoebe. This 132-mile wide body was interesting in that it appeared to be a body captured from the outer solar system, as opposed to one incubated near Saturn—the fact that its orbit is backward (or retrograde) and out of plane with many of the other moons supports this conclusion. It is hypothesized to have a large volume of water ice below its surface. Phoebe also has its own faint, diffuse ring, thought to be a byproduct of other bodies smashing into it over the course of its life.

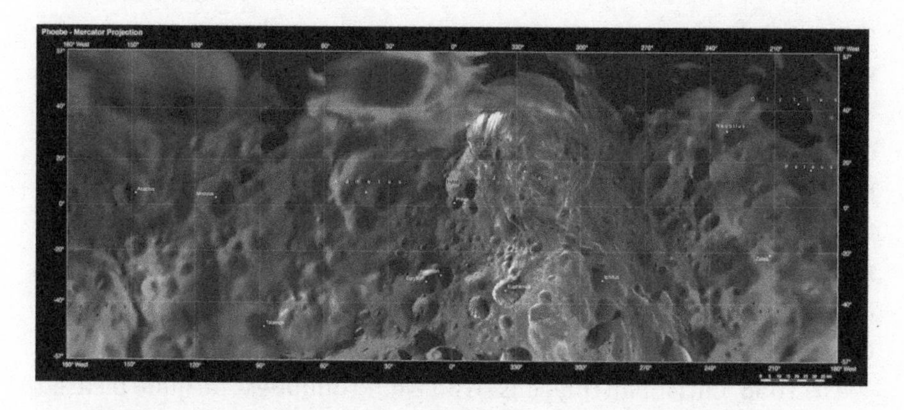

Fig. 32.2. A Mercator map of Phoebe's surface built from images captured by Cassini's single pass of the moon (orbital mechanics proscribed further study). Its fast rotation of about nine hours allowed Cassini to image the entire surface as it sailed by at a distance of about two hundred miles. Image from NASA/JPL-Caltech.

After this brief reconnaissance, Cassini moved on to begin its long relationship with Saturn and its other moons.

Then Cassini made a daring maneuver. After long and complex calculations of the orbital dynamics of the Saturnian system—which still involved a bit of guesswork—the spacecraft was oriented toward a wide gap in Saturn's rings for a fly through. It was risky, but why come to a baseball game if you're not going to swing for the bleachers? On July 1, 2004, the probe passed through a gap between Saturn's F and G rings. That isn't quite as dangerous as it sounds—these rings are thin and sparsely populated, and the gap was sufficiently wide to not be considered much of a risk to the spacecraft. In fact, the F and G rings are so tenuous that they were only discovered when the Pioneer and Voyager images came back and showed the thin, thready structures.

Before we continue, I should give you a bit of a real estate agent's tour of the rings. They were first spotted by Galileo through his small telescope, and he initially thought them to be additional planetary bodies, writing, "The planet Saturn is not alone, but is composed of three, which almost touch one another and never move nor change with respect to one another. They are arranged in a line parallel to the zodiac, and the middle one is about three times the size of the lateral ones."[6] Through his tiny telescope, the edges of the rings looked like two more smaller planets. Observing Saturn two years later, he noted that the objects he thought were two more bodies had vanished—the rings were edge on as seen from Earth, but of course he could not know that. In the following year, they became ever-so-slightly visible again—this must have been confusing for the valiant proto-astronomer.

In 1655, Christiaan Huygens (who would doubtless be quite pleased to have a probe named after him for a moon he discovered) used an improved telescope to observe that the apparition was in fact a ring of some kind surrounding the planet. He noted that the planet was "surrounded by a thin, flat, ring, nowhere touching, inclined to the ecliptic."[7] Twenty years later, Giovanni Cassini saw the ring divisions (one of which is named after him, the Cassini Division) and realized that the rings were separate structures in a flat plane around the distant world.

Over the next two hundred years, scientists like Pierre-Simon Laplace and James Clerk Maxwell intuited that the rings must be made of particles—they could not be solids or liquids as others had theorized; mathematical calculations showed that such structures would not be stable in orbits around a large planet.

As the decades sped by, other astronomers used ever-larger instruments to study Saturn's rings, but the real cornucopia of data prior to the Cassini mission came from the Pioneer and Voyager spacecraft as they sped past the planet, with Voyager 2 passing at about 23,000 miles.[8] Only then did the real structure and composition of the rings become entrenched in the science books.

Before we get too far, let's clear up one thing: the rings are out of sequence. They were named in the order they were discovered, which makes some sense, but not sequentially from the inside out, which would have made more sense. It's a bit like trying to find an address in Tokyo, where homes are numbered in the order they were built (if you've never experienced it, finding a home in an older neighborhood there can be a real chore).

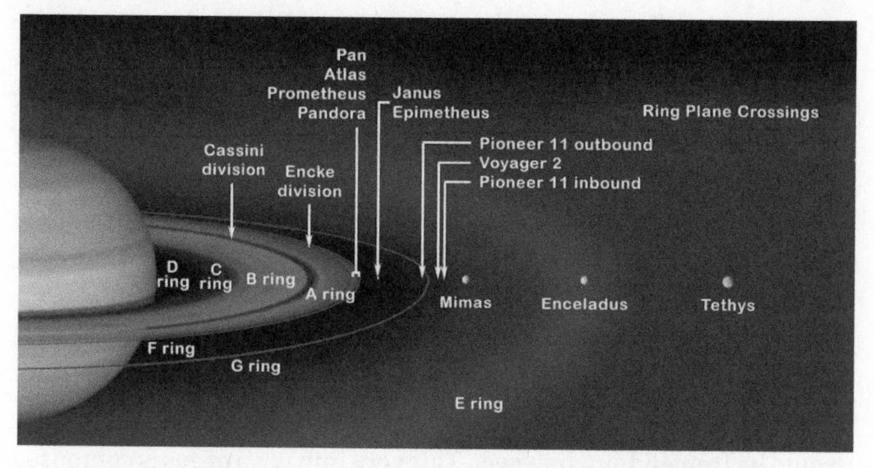

Fig. 32.3. Saturn's rings and nearby moons. The letter identifications of the rings are out of sequence because they were labeled as they were discovered—for example, the F and G rings were spotted by Pioneer and Voyager, centuries later than many others were observed. Image from NASA/JPL-Caltech/SSI.

Despite their stunning beauty and profound appearance from way across the solar system, about 1.3 billion miles from Earth at the closest, those magnificent rings are only about thirty feet thick on average—not much more than the length of a school bus. They vary in width substantially, in few places they are even close to a mile thick (and there are a few exceptions that range up to many miles). They are composed almost entirely of water ice, with a smattering of sand and mineral dust—pity the missing moon that gave up its existence so that we could have the spectacular rings to look at. And, speaking of moons, if all the mass of those huge rings were packed into a ball, it would only amount to a medium-sized moon. Personally, I like them a lot better the way they are—Saturn has plenty of other moons to gawk at.

The gaps in the rings that add to their majesty are caused by gravitational interactions of some of those moons, and in some cases are a more direct result of a moon. The latter are Saturn's "shepherding moons," such as Pan, Saturn's innermost satellite. Pan is shaped like a twenty-one-mile-wide walnut, and carves its way along Saturn's A ring, in what is called the Encke Gap, sweeping aside or smashing into any errant ice that it hasn't already bulldozed in Pan's long life. Variations in its orbital distance from Saturn have helped Pan to sweep clean a 201-mile-wide gap. The distinctive ridge surrounding its middle is thought to be a result of all the junk it's swept up over many millions of years—Pan is the trash collector of the Encke Gap.

Other gaps, such as the massive Cassini Division, are formed by gravitational interactions by moons beyond the rings, such as Mimas, with the particles that make up the rings.

The rings even have a very slight atmosphere caused by gasses escaping the mostly water ice that comprises them, though not a sufficiently thick one to cause you to want to have dinner there. The view would be tremendous, however. This very thin wisp of gas is composed of oxygen and molecular hydrogen.

The mechanism of the formation of Saturn's rings is uncertain—they are either an unformed moon left over from the birth of Saturn itself, or one (or more) satellites that were smashed up during a particularly violent phase of the early solar system about four billion years ago. Other theories peg the age of the rings closer to 100 million years ago—a blink of the eye in solar system terms—and cite the brightness and observed purity of the water ice as evidence. Regardless of how they formed, the outer A and B rings appear to be the densest, with the others being comprised of dust—some of them are almost the consistency of smoke.

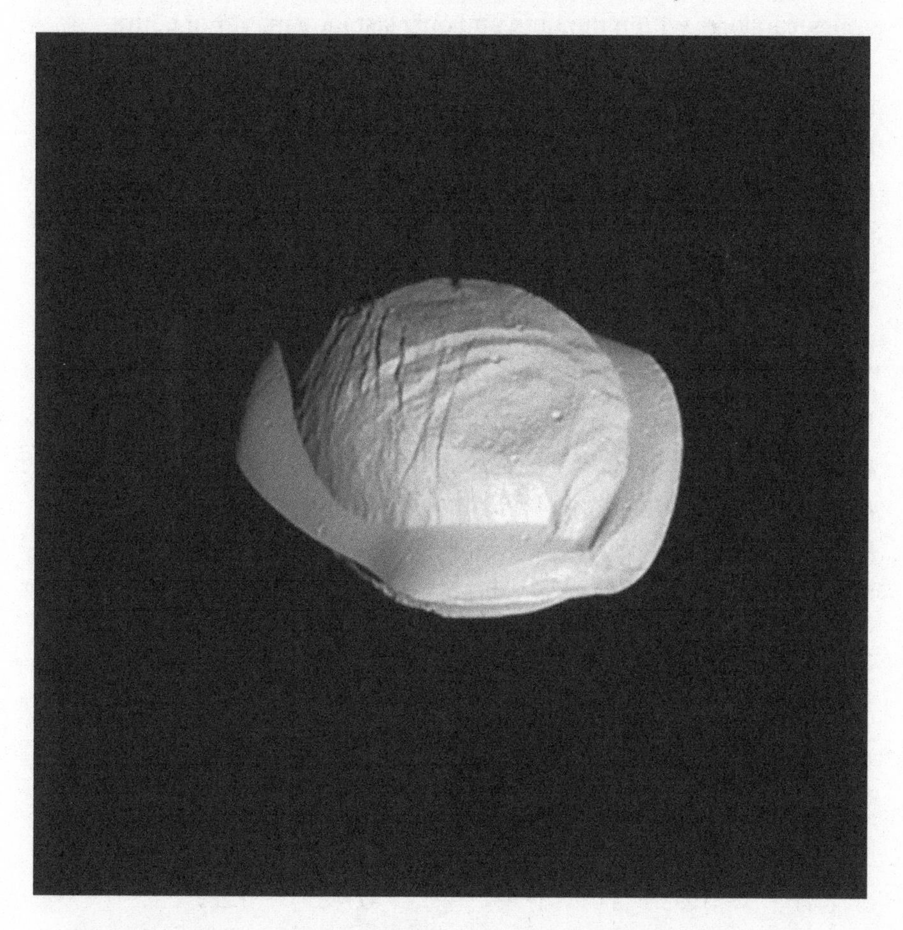

Fig. 32.4. Pan is one of the solar system's weirdest moons, and has alternately been described as resembling a walnut, a ravioli, or an empanada. The distinctive ridge is thought to have been formed by the moon's gathering of material as it sweeps along the Encke Gap. Image from NASA/JPL-Caltech/SSI.

Adding to the mystery and complexity of the rings are the spokes that periodically form and are best viewed in movies created from the many still frames returned by Cassini. These appear as dark smudges when backlit, and move around over time. The spokes are thought to be a result of electrostatic repulsion caused by Saturn's intense magnetosphere and its interactions with the denser rings. These features come and go in cycles, can form within minutes, and only last for a matter of hours.

One more interesting facet of the rings is what appears almost to be a foam, which casts shadows across some of the ring edges. One such region is along the Keeler Gap, which is only about twenty-six miles across in the A ring. A tiny moon called Daphnis, only about five miles wide, scours this area. As Daphnis orbits Saturn, it wobbles a bit, and causes a disruption along the edges of the surrounding rings. With the proper lighting angle, this disruption shows up as shadows cast across the nearby ring planes. These disruptions are at most only about a mile high.

Finally, we have the faint ring caused by the moon Enceladus. This fascinating world is one of Saturn's larger moons, at about 318 miles across at its widest, and has a subsurface ocean under its icy crust. Apparently geologically active, Enceladus has frequent geyser-like eruptions from its surface that form the very faint E ring that extends all the way from Enceladus's orbit out beyond that of tiny Rhea.

For those of you who enjoy such things, a researcher at the Goddard Space Flight Center in Maryland has put together a nifty chart of the ring widths.[9]

Widths of the Rings.

Ring	Saturn radii	Kilometers	Miles	Earth diameters
D ring	0.126	7,594	4,719	0.595
C ring	0.288	17,357	10,785	1.361
B ring	0.424	25,554	15,878	2.003
A ring	0.242	14,585	9,063	1.143
F ring	-	-	-	-
G ring	0.08	4,821	2,996	0.378
E ring	5	301,340	187,244	23.623

(The F ring data is blank because it's a thready, spindly thing only a couple of hundred miles wide that weaves around and is presumably too small and irregular to merit inclusion in this table.)

That's about enough of the realtor's extolling of the view out the windows . . . "Oh, just look at that G ring, will you? Scrumptious." Let's get to the meat of the property—location. And that location is around one fascinating planet.

Devoid of its rings, Saturn would play a mousy second to Jupiter. The gaseous surface is far less dramatic than its larger neighbor, and in uncorrected photos appears to be a rather bland shade of tan with weak variations between the bands. But that bland exterior belies the violence underneath, as was dramatically illustrated in late 2006 when Cassini observed a storm in the south polar region of the planet, a hurricane-like cyclonic disturbance about 5,000 miles across. This disturbance appeared to be stationary, with winds up to 350 miles per hour. Then, in 2010, another storm, one about halfway between the north pole and the equator of the planet, kicked up—and what a sight it was. The storm was spawned by a feature called the Great White Spot, which forms every thirty years or so, and on this occasion was briefly larger than Jupiter's Great Red Spot. Temperatures in the area rose briefly, at one point topping 150 degrees Fahrenheit, and an associated increase in radio noise and electrical activity was charted by Cassini. These storms are apparently caused by seasonal cooling of the atmosphere, during which water (which exists in more abundance here than on Jupiter) falls to lower altitudes, leaving a low pressure area of hydrogen and helium that cools and densifies (though, as noted above, periodic discharges can also cause heating). This can cause storms such as the one observed. The most spectacular aspect from a visual standpoint was that the spot became an elongated series of interconnected blotches that looked like someone had dribbled white paint while carrying a leaky bucket above the cloud tops—the spots were a continuous streak that ended up covering much of one hemisphere. In 1990, such an occurrence in fact encircled the northern latitudes of the planet. On an otherwise beige world, the disruption was stunning.

Throughout the mission, Cassini flew past the moons of Saturn, some on multiple occasions, to study these fascinating bodies, the most noteworthy of which were Titan and Enceladus. First, Enceladus.

Enceladus was targeted for twenty-two passes throughout Cassini's lifetime. It is the brightest moon in the solar system in terms of reflectivity, and its geologically complex surface demanded further study. The discovery of the geyser-like plumes emanating from the world also intrigued—this was a moon that deserved increased attention.

Regarding that brightness: it turned out that both the surface of the moon and the material that made up the E ring originated from Enceladus's interior—those plumes were responsible for decorating both the moon and its environs with a coating of highly reflective ice. For a smaller moon—it's only about 310 miles across—it's a very busy place.

The geysers erupt violently, with the spray moving at about 800 miles per hour, extending hundreds of miles into space. And while most of this appears to be water, microscopic particles of silica were spotted in the adjoining E ring, indicative of high temperature interaction of water and rock (about 200 degrees Fahrenheit or higher). This indicated hydrothermal vents on the seabed of Enceladus, far below the icy surface. And where there's water, heat, and minerals, there could be life—the extremophiles found on the seabed of earth at hot vents, where mineral-rich water can come spewing out at upward of 870 degrees, are proof of that. It was a turning point in the study of the outer solar system. As Spilker put it, "Enceladus discoveries have changed the direction of planetary science ... Multiple discoveries have increased our understanding of Enceladus, including the plume venting from its south pole; hydrocarbons in the plume; a global, salty ocean and hydrothermal vents on the seafloor. They all point to the possibility of a habitable ocean world well beyond Earth's habitable zone. Planetary scientists now have Enceladus to consider as a possible habitat for life."[10]

How did they get all this detail you might ask? Well, like any true hands-on geologist, Cassini did some close up and personal fieldwork. During a couple of the flybys, one as close as thirty miles above the moon's surface, Cassini flew right through the plumes and sampled the vapor. It detected a mix of volatile gases, water vapor, carbon dioxide, and carbon monoxide, as well as organic materials. The density of organic materials was about twenty times denser than expected. The subsurface ocean appears to be about six miles deep, under an icy crust some nineteen to twenty-five miles thick.

It became clear during the course of the mission that the cracked terrain on Enceladus was in slow but continuous motion, stretching, buckling, and cracking. Upwelling water often freezes over the wounds, temporarily "healing" them.

And then there was Titan, the Metropolitan Paris to Saturn's France. By far the largest of the planet's moons, it is also in many ways the most compelling (and even has smog to challenge Parisian levels as well). In fact, you would not be misguided to call Titan "Smogworld" if you were contemplating writing Kevin Costner's newest screenplay. And with a diameter of 3,200 miles, Titan is second only to Ganymede in the moon-sizing contest, and both are larger than sunny Mercury.

Titan is unique in that it is the only moon in the solar system with a dense atmosphere that also has bodies of standing liquid on the surface. But before you dive in, realize that these lakes are more like gasoline than the local community pool, filled with liquid methane and ethane at temperatures hundreds of degrees below zero degrees Fahrenheit. Ligeia Mare, the second largest body of liquid on Titan, was confirmed to be composed of liquid methane in 2016 via Cassini radar soundings. These hydrocarbon liquids also form rivers and streams crisscrossing the unearthly surface. The copious amounts of water on Titan are frozen masses, and the surface is laden with what appear to be sand dunes but are actually formed of granules of organic compounds, looking like spent coffee grounds. These granules form high in the atmosphere, possibly via the interaction of methane and sunlight, and fall like snow to form dunes around the moon's equatorial regions. Titan's surface is relatively free of the cratered scarring of other moons and planets, as the weathering continuously erodes and reforms the terrain.

Titan's atmosphere is about 45 percent thicker than Earth's, and moving through it would feel more like swimming in water than an afternoon stroll. With Titan's lower gravity, that atmosphere extends about 370 miles up, much farther than Earth's. It is composed largely of nitrogen, at about 95 percent, with the rest made up primarily of methane. This methane is what forms the granules that litter the equator—if you were to stand still on this part of Titan (so that you didn't notice the density of the air), you might think you were somewhere in a smoggy Sahara—just don't take a breath.

On December 5, 2004, the Huygens probe detached from Cassini and went its own way, bound for a rendezvous with Titan. About a year later, on January 14, 2005, Huygens plunged into Titan's atmosphere. It slowly descended through the thick envelope of nitrogen, eventually coming to rest after a two-and-a-half hour descent, transmitting to the Cassini spacecraft all the while.

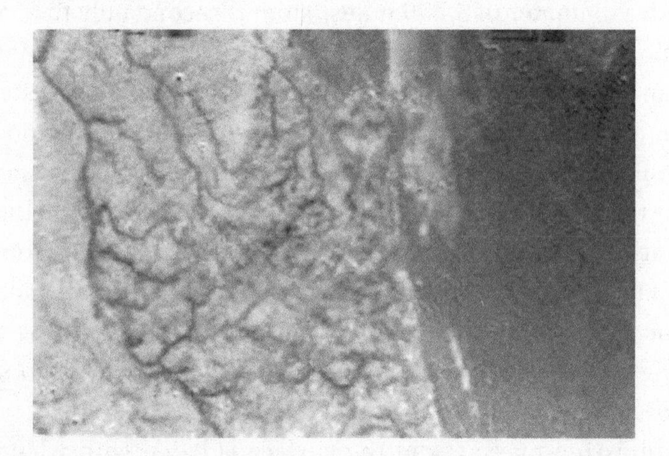

Fig. 32.5. The first image from the Huygens drop probe shows what are presumed to be drainage channels and a possible shoreline into which they flow. It might be a beach, but not one you'd enjoy. Image from ESA/NASA/University of Arizona.

When the probe hit the surface, it slid about a foot then came to rest. The rocky area it landed in was not far from a dry, former shoreline and had the consistency of packed snow. The rocks and pebbles nearby appear to be worn, suggesting that they were eroded via transport by a liquid.

The local temperature was about -290 degrees Fahrenheit, with high humidity (of methane, that is). Dew of some kind was observed at the landing site. Light levels were about equivalent to late sunset on Earth, with a small sun shining wanly through the dense orange haze.

Measurements from Cassini indicated that there is a large subsurface ocean of water on Titan, probably briny and laced with ammonia, as deep as fifty miles below the surface. If warm, this could be one more place in the solar system where life could have formed—it is even possible that some kind of life exists on the surface, though it would be in the form of an extremophile. There is some evidence of tectonic activity at work.

The observational period on the surface was brief, as most of the Huygens's three hours of battery life was used up during its descent, leaving about thirty minutes of power remaining once the probe was on the surface. But that time was used well, with a vast amount of data transmitted along with some imagery. It is an intriguing world, and we will surely be visiting again.

Cassini was a magnificent mission, unveiling countless secrets just hinted at by the Voyager flybys. But all machinery wears out, and only a limited amount of fuel can be carried along for the ride. After thirteen years at Saturn, and almost two decades in space, by 2017 Cassini was tired, radiation-battered, and low on fuel. It was time to retire the aging machine.

Since 2010, mission planners had been considering this moment. As they had with Galileo, the team considered the options. They did not want to risk contaminating any of the icy, ocean-bearing moons with a dirty spacecraft (in microbial terms, that is). Both Enceladus and Titan were high on the list for future astrobiological missions, and thus needed to be protected. An accidental encounter on any moon with a tumbling, derelict, plutonium-carrying spacecraft would not serve anyone's purposes. The safest bet was to drive Cassini into Saturn itself, where it would be vaporized and do no harm. But before that, while they still had maneuvering fuel, the time was right to take some of those big chances that nobody in their right mind would do during an ongoing mission.

The final twenty-two orbits of the Saturnian system were planned out to take place over the last five months of Cassini's life. They would take the probe through the gap between the innermost rings and the planet, where no machine had ever dared venture. This would allow the scientists to measure the mass of the inner rings, and to address their origin and age if possible.

Each orbit took just under seven days, and each brought the spacecraft a bit closer to the planet's cloud tops. Finally, on September 11, 2017, the spacecraft took a last spin past Titan, a close pass that altered Cassini's trajectory sufficiently to induce one final, fatal plunge into Saturn.

On September 14, Cassini began its final eleven-hour downlink of data from its onboard tape recorder. For most of its passes near the rings of Saturn, the spacecraft had reoriented itself dish-forward to protect the body of the spacecraft from the impacts of tiny particles, then reoriented itself back into position to communicate with Earth. But for the last 14.5 hours of the mission the antenna would remain locked onto Earth, transmitting continuously. JPL wanted to squeeze every last bit of possible science out of the mission. While the spacecraft had relied on quickly spinning reaction wheels, a type of gyroscope, for maintaining the proper orientation throughout its two decades in space, it had fuel to spare for the time it had left. The maneuvering jets were armed, and they would be used for the remainder of the mission to overcome the slowly increasing drag forces of Saturn's atmosphere and keep the radio dish pointed Earthward.

On September 15, I was back up at JPL to watch the last act of Cassini's mission. At just past midnight Pacific Daylight Time, the downlink switched from flight recorder data to real-time—the science teams didn't want to miss even a moment of what Cassini observed in its final hours. The recorder wasn't needed anymore, in any case—there would not be a second chance to re-send any data.

At about 4:54 a.m., Pasadena time, Cassini encountered Saturn's upper atmosphere (all these times were delayed by almost ninety minutes from when they actually occurred due to the long radio signal delay from Saturn). The spacecraft was traveling at about 77,000 miles per hour now, and flight controllers had long since sent up a command telling the spacecraft to turn on its maneuvering thrusters at that time. The small rockets fired up to about ten percent, just enough to keep it properly oriented to keep transmitting. Cassini was still 1,200 miles above the clouds. About a minute later, the thrusters throttled up to 100 percent—the spacecraft was starting to feel the effects of the thin atmosphere—it had never flown anywhere near this close to Saturn's clouds before—and it was experiencing some heating via friction.

All along, we'd been watching the progress of the end of the mission on the big screen at the front of Von Karman Auditorium. Some expected dramatic video, showing ominous tan clouds rushing ever closer—but they were to be disappointed. All that had been displayed were a few computer animations earlier in the evening, describing what would be happening. Then, for the last forty-five minutes or so, there was just a graph, like an old cathode-ray tube oscilloscope (the kind you may have seen in 1950s science fiction movies). It was a graph of the strength of Cassini's radio signal, and it looked very much like the heartbeat monitor of a dying patient in a hospital.

The auditorium was nearly full, with the press joining people who had devoted a decade or more of their lives to this mission, and who were willing to come out in the wee hours of the morning to watch it end. Expressions ranged from engaged curiosity, to wonder, and the closer the end came, sadness.

All we saw was a graph of the radio signal coming in from Cassini, represented by a small green spike at the top of a dome-shaped graph plot. From top to bottom, as that central spike diminished, we knew that well over a billion miles away, Cassini was being buffeted, scorched, nudged off axis, and finally torn apart in its final plunge into Saturn when the line went flat.

The room was quiet for a few minutes. The clock read just a bit before 5:00 a.m. It was over, and gradually applause started up. There was hugging, a few tears, and even more smiles. A press conference was held a bit later, in which team leaders waxed enthusiastic about an incredible mission that had consumed years, and in some cases, decades, of their careers.

Fig. 32.6. Julie Webster, a chief engineer of the Cassini mission, to right, hugs a team member in mission control shortly after the signal from the spacecraft was lost. Image from NASA.

So what's next at Saturn? Spilker had some thoughts. "We've put together a proposal . . . to go back to Enceladus with the kinds of instruments that you would need to address the questions about the habitability and is there life in the ocean of Enceladus . . . The mission's called Enceladus Life Finder."[11]

It's just a proposal for now, but what a grand mission that would be. Enceladus Life Finder, or ELF for short (I can't wait to see the mission patch for that one) would be an orbiter, not unlike the Europa Clipper, that would circle Enceladus at a low altitude, specifically targeting geyser plumes through which to fly. During each pass it would capture some of the vapor from the plume and run it through instruments that would include dual mass spectrometers, which would give a chemical breakdown of the plume, and also seek amino acids, isotopic ratios and the amounts of any methane present. ELF will be a true astrobiology mission, intended to look for organic molecules and biogenic signatures, the first steps toward finding life beyond Earth.

CHAPTER 33

FLASH FORWARD: DIVING ON TITAN

What kind of spacecraft do you send to the only other place in our solar system that has standing oceans? Now, let's complicate that request with the fact that this world is Titan, and one of the choice targets, Kraken Mare, which is larger than some of the United States' Great Lakes, appears to be filled with liquid methane and some ethane. It is about six hundred miles wide, almost as wide as Texas, with a depth of over a thousand feet in some places. Add to the equation that the methane sea is about -297 degrees Fahrenheit in temperature. And at the bottom? A rich organic sludge is hypothesized . . . and a sample of that could be a scientific windfall.

So how do you explore a frigid, deep, hydrocarbon lake? With a submarine, of course. A number of such missions have been proposed, and one of the most daring came from a NASA program called NIAC (NASA Innovative Advanced Concepts), which funds some of the most out-of-the-box thinking in space exploration.

There have been a number of proposals, but one of the most exciting envisions an autonomous submarine that looks like a torpedo with a grid-like fin on the back. The 3,000-pound submersible would be powered by nuclear power plants—not reactors as are used in Navy submarines, but radioisotopic generators as have been used on many of NASA's deep-space explorers. The heat from the plutonium fuel would also be circulated through the machinery to keep the sub warm in the otherwise crippling cold.

Fig. 33.1. Kraken Mare, one of the largest of Titan's hydrocarbon lakes. At nearly 300 degrees Fahrenheit below zero, and with depths of up to a thousand feet, navigating this forbidding extraterrestrial sea would be an unprecedented challenge. Image from NASA/JPL-Caltech/ASI/USGS.

Onboard instrumentation would be capable of scouting not just the depths of the hydrocarbon sea but also the surface. On the surface, sensors would measure the weather and light levels, and would be capable of imaging the surrounding waves and nearby shoreline. When submerged, the sub would measure the chemical properties of the ocean, and would likely carry a sampling arm to retrieve items of interest—primarily the organic ooze that is theorized to be at the bottom of these seas—for analysis in an onboard laboratory. The probe would also have a depth ranging device and side-scan sonar, a technology that creates a fan-shaped sonar beam that can be interpreted to create 3-D maps of the ocean floor.

Mission cycles would involve eight-hour dives followed by sixteen or more hours on the surface, when the submersible would conduct more observations and transmit its stored data to an orbiter, which would record it for transmission to Earth. This would also be the time that any incoming instructions from home would be relayed to the submarine, which must be surfaced to communicate with an orbiter, or possibly directly with Earth. Operation would be highly automated due to the long periods of blackout when submerged, which, when combined with the long transmission times to and from Earth, require a high degree of autonomy.

NIAC studies are an opportunity for engineers and scientists of different stripes—not all from NASA, and many who have never even worked in space-related projects—to come together and conceive innovative project designs. Experts in underwater propulsion, submarine hull design, atomic propulsion concepts, and even torpedo design were consulted.

Exotic ballast systems will be required to control the buoyancy of the submarine. The techniques that work in Earth's watery oceans would not apply on Titan. Normally, a submarine ingests seawater to sink, then displaces it with compressed air to surface. But on Titan, that "air"—the atmosphere—is nitrogen, which will liquefy below a certain depth and pressure at these temperatures. As the study's lead put it, "The temperatures there are so low, when you get down to a pressure of about 100 meters or so, the nitrogen collapses back into a liquid. . . . Our solution is to bring neon [filled] tanks and have a piston system, and you basically pressurize with neon and push out the ballast methane/ethane so that you can float."[1] Neon won't compress into a liquid at the temperatures found in Titan's oceans.

The nitrogen in Titan's atmosphere poses other potential challenges as well. If a sufficient amount of it has dissolved into the sea, which seems likely, it could turn back into gaseous bubbles when churned by a propeller, or when heated by the submersible's excess heat. This could result in a cloud of fizzy bubbles surrounding the submarine, which would hamper its observations, and even negate the effect of the propellers—it would be like using a boat propeller in air. The solution to this problem is a work in progress.

A number of ways of delivering the submarine to Titan were investigated, but one stood out for both originality and economy. If sized properly, a submersible could be packed inside the payload bay of the US Air Force's X-37B, an automated mini-shuttle that has been conducting missions successfully in Earth orbit for years. The X-37B would be lofted into orbit with a conventional rocket, then boosted toward Saturn with a powerful upper stage. When it reached Titan, the X-37B would glide into the atmosphere, just as it does in Earth, and eject the submersible above the lake. The X-37B would then ditch and sink, leaving the submersible to continue on its mission of exploration.

The X-37B is currently flying regularly, and UAVs—Underwater Autonomous Vehicles—are in use all over the planet. The key to success on Titan is to blend these technologies, develop alternate methods to deal with the exotic environment, and engineer into it a high level of autonomy.

If NASA has its way, we could be exploring the seas of Titan within twenty years.

Chapter 34

A RETURN TO MARS

In the next few years, the Mars rover Curiosity will have a twin on the planet, much as the Mars Exploration Rover Opportunity did when its sibling Spirit was still roving half a planet away. The Mars 2020 rover (which does not yet have a formal name) is scheduled to launch in the summer of 2020 on an Atlas V rocket and arrive at the red planet in early 2021. With Opportunity apparently down for the count, we will once again have two rovers working on Mars.

The rover is an overall clone of the Curiosity design that worked so well in 2012, with about 85 percent of it being identical. It will even use some parts left over from the Curiosity mission, including a heat shield, to save money. But there are differences deep inside. While it uses the same overall chassis design, and will use the sky crane system to land on the planet, this is a true astrobiology mission, and the instrumentation within is tailored to a search for living things, past or present.

This search for life is where Mars 2020 parts company with Curiosity. That rover's primary toolset was split between instruments on its robotic arm and those within the chassis. The small sample drill ground up bits of rock into a powder and delivered that powder to investigative instruments onboard. Mars 2020 will do things a bit differently.

In a nutshell, the onboard lab has been moved to the end of the robotic arm. Rather than gathering samples and moving them into an onboard lab, Mars 2020 will use instruments mounted on the arm to study areas of interest in microscopic detail.

The downside of the approach that Curiosity used is that some of the context of the sample was lost—when the rover gathered a bit of Mars and analyzed it, in the process of collecting and moving the sample some of the details would get muddled.

Here's how Ken Farley, the project's chief scientist, put it: "Previous rovers used sophisticated analytic instruments and prepared rock and soil specimens for analysis on board the rover itself. . . . Conceptually, Mars 2020 marks a transition from missions in which sampling guided exploration to one where exploration guides sampling. In other words, the rover's scientific instruments will observe the surrounding terrain and provide the critical context for choosing where samples will be collected. Ultimately, this context will also be used to interpret the samples."[1] In the planetary exploration game, context—the analysis of a sample where it resides—can be everything.

As with Curiosity, the camera mast at the front of the rover carries two main instruments that will guide the rover to areas of general interest. Mastcam-Z is a stereo camera for helping to guide the rover, image the surrounding terrain, and identify and record areas of interest. This new camera incorporates something that the geologists have craved and the engineers feared for almost two decades: a zoom lens. Long considered too bulky and risky—"It's got moving parts! Those can fail!"—a zoom has finally been designed that is considered reliable enough to travel to Mars. The second instrument is called SuperCam, which is an evolved version of a similar instrument on Curiosity. SuperCam sports a laser that is powerful enough to vaporize spots of rock or soil from many yards away, then analyzes the light given off by the incandescent flash with spectrometers. This approach was very successful on Curiosity, allowing targets of interest to be evaluated from many yards away.

In total, Mars 2020 will have twenty-three cameras aboard, for both navigation and investigation. The rover will make macro-level decisions using the cameras and spectrometers on the mast; Mars 2020 can then "move in for the kill," and get microscopic. On the end of the rover's arm are more instruments, and these will move far closer—

instead of looking at an element of the wider landscape, or grabbing a sample from an area the size of, say, a postage stamp, Mars 2020 will now be looking at individual grains of sand or specks on a rock.

One of these instruments is called PIXL, for Planetary Instrument for X-ray Lithochemistry. PIXL places the instrument's capability out onto the arm, where it can use X-rays to excite minerals in-situ, directly at the point of interest, and identify their composition. Another instrument on the arm, called SHERLOC—for Scanning Habitable Environments with Raman and Luminescence for Organics and Chemicals (this won't be winning any arcane NASA awards for brevity in naming)—uses ultraviolet Raman and fluorescence spectroscopy to map out any organic matter found on the sample. Raman spectroscopy uses monochromatic sources of light to excite a sample, and specific characteristics of that sample can be identified at a molecular level—it is very exact, and has nothing to do with instant noodles ("Raman" confuses the hell out of many spellcheckers).

To complete our survey of the Mars 2020 robotic arm, SHERLOC also includes a very powerful microscopic camera. This will help to identify tiny crystalline structures in rocks and also provide a microscopic visual context for the sample being investigated.

There are also other instruments mounted on the main chassis and around the perimeter of the rover. RIMFAX, the Radar Imager for Mars's Subsurface Experiment, will send radar waves into the ground to seek water and ice below the surface. It can measure with an accuracy as precise as three inches and "see" as deep as about thirty feet into the ground. It will run more or less continuously as the rover drives along, seeking underground deposits of water in any state.

A weather-detecting instrument will be present, as has been carried on all previous Mars rovers. Called MEDA (Mars Environmental Dynamics Analyzer), this instrument will measure temperature, wind, humidity, and several other characteristics of the Martian environment. Finally, our good friend MOXIE (the little oxygen factory) will be up front (see chapter 14).

That's the sweep of the main instrumentation onboard Mars 2020.

But let's get back to the conversation about life. I said this was to be an astrobiology mission—how exactly will this quest be accomplished?

Much has been learned since the days of the Viking landers, and Mars 2020 will be using very different devices to seek life. Our education about how to seek possible life on Mars has come in a variety of forms, primarily through an enhanced understanding of the Martian environment and surface chemistry, as well as via new discoveries about life here on Earth, which can take much more extreme forms than we knew about in the 1960s and 1970s.

The systems that will be used in this search are quite sophisticated and have evolved even since Curiosity was launched. There are two major steps to identify possible biological activity on Mars. The first is to search for organics with the instruments described above, in particular with SHERLOC. The second is to select key areas for drilling, collect a core sample, and then send it home to Earth for analysis. But let's not get ahead of ourselves here—first, let's take a look at the new drill.

Curiosity's drill looked much like a traditional, home-tool-shelf drill bit combined with a flat screwdriver tip. It was designed to dig into dried mud, sandstone, and rock via a combination of twisting and hammering, and to deliver just a tiny bit of powdered sample to an attached container out on the end of the robotic arm. That sample powder was then delivered to the instruments inside the rover's main body by the arm. Mars 2020 uses a true core sample drill, a hollow tube just under six inches long and about three-quarters of an inch wide. These tubes are made of titanium, have a tough tungsten-carbide drill bit, and once they drill into rock or soil, are brought back toward the rover, hermetically sealed with a metal cap, and stowed on the rover or used to provide a half-ounce sample for the onboard analytics lab. The rover will carry as many as forty-three tubes—thirty-one for samples, seven "witness" tubes, and six spares. The witness tubes carry control samples for comparison—little bits of Earthly contamination—that will be exposed to the same conditions as the Martian samples. Then, when the samples are brought back to Earth for analysis, the soil and rock can be compared to the "witness" samples, thereby minimizing the changes of a false identification—that item of interest you're seeing might not be Martian life, but contamination from Earth that went along for the ride and was affected by exposure to a deep-space environment. The same could be true of certain chemicals or organic molecules, so having "controls" is critical. In fact, one of the witness tubes will even be left open to space as the rover flies to Mars to collect any contamination that might "outgas" or come off of the rover as it flies—you cannot be too detail oriented in the astrobiology business.

But back to those samples. After cores are drilled and stored, they will be left on the Martian surface for later retrieval. The locations of these cached samples will be carefully recorded. Years later, a "retriever rover" will be sent to pick them up and deliver them to another lander, which will have a sample-return rocket (the Mars Ascent Vehicle, or MAV), with which the samples will be sent back to Earth (see chapter 27). While the retrieval mission is still in the planning stages, and not yet funded by NASA, the caching of samples by the Mars 2020 rover should apply some pressure to NASA and Congress to get that next mission up and running. It would be a shame to leave pristine samples of Martian rock and soil sitting on the surface for too many years before retrieving them. The tubes will be coated with a special aluminum oxide to reflect most of the sunlight that could overheat them, which will keep them acceptably cool while they await retrieval.

When these samples are eventually returned to Earth, they will either come directly back to our planet via a robotic lander, or they will be sent to a lunar-orbiting space station where astronauts will begin to perform a first round of analysis.

In any case, if something that indicates a possible form of life—either fossilized or current—is found in a sample, scientists all over the world will a) be excited, and then b) try to prove that it ain't so. That's how science works ... attempt to disprove before accepting. Hence the witness tubes—you just can't be too careful.

Another part of that caution is an almost hysterical attention to sterilization of the spacecraft.

When Curiosity flew, it was cleaned as well as possible, but since it was not a life science mission per se, but one intended to seek out habitable soil chemistry in the Martian environment, sterilization was not critical—just highly desirable. In the end, some last-minute fix-its with the drill probably introduced some contamination into the system just prior to launch, but so far that has not proved to be an issue on that mission.

Mars 2020 will be different. Any contamination could be a critical issue, since scientists will be looking for organics and other substances in the parts-per-billion range. That's teeny-tiny, and you've got to have a clean spacecraft. Until now, the sterilization protocols for the Viking landers of the 1970s have been the golden standard—repeated cleanings with chemicals designed to kill everything other than their human handlers. This was a level of careful handling that had never been accorded a spacecraft before, and, finally, after enclosing the landers inside their hermetically sealed aeroshells, NASA subjected the probes to a high-temperature baking at 233 degrees Fahrenheit for thirty hours. That's thorough, and resulted in a "bioburden," or bacteriological count, of less than three hundred spores per square meter of spacecraft surfaces. This has been the highest level of sterilization ever attempted—it has never been exceeded. Mars 2020 will not be subjected to measures quite this severe—the onboard electronics are too sensitive for that kind of baking—but it will be heated and scrubbed to within an inch of its tolerance, and any devices that touch the surface of Mars will receive special attention and be hermetically sealed prior to launch.

All that said, there are still limits as to where the rover is allowed to explore. NASA has a special office, called Planetary Protection, that curates the rules about what spacecraft can go where—their goal is to protect anything that might *possibly* be alive, anywhere off Earth. Any part of a planet or moon that has liquid water and might be a hotbed for biological activity is currently off-limits to JPL's exploration machines. This causes some conflict between the explorers and the bio-cops, as you might imagine, with some astrobiology-types saying that Mars is so toxic to life on its surface that any contaminating organisms brought along with a rover would be dead within hours. The planet is exposed to far more radiation from space than Earth, due to its lack of a dense atmosphere and its almost nonexistent magnetic field. Earth has both, which protect our planet from most space-borne radiation and make our microbes here far weaker than they would need to be to survive on Mars or elsewhere. Further, the soil chemistry on Mars is also thought to be hostile to life, so these folks feel that the protections are too severe, impeding science. They further point out that the moment humans set foot on Mars, all bets are off anyway, as those astronauts will carry germs with them no matter how careful they are—even on the outside of their space suits—since people are germ-breeding machines, basically just walking petri dishes for little critters.

But for now, the Planetary Protection people are holding the line—nothing goes near any source of "wet" water anywhere off Earth until NASA can ascertain how to *perfectly* sterilize the machine, a nearly futile endeavor with current methods. It will likely be some time before this is settled. For Mars 2020, the goal is no more than forty parts per billion of organic carbon in any sample, for example. That's pretty tight as it is.

But Mars 2020 will collect core samples from anywhere it is allowed to explore, and when these are eventually returned Earthward for analysis, scientists will hunt for amino acids, precursors of proteins, and other complex organic compounds. They will also analyze the isotopic ratios in key molecules—some are clear indicators of biological processes. It's still a bit of an inexact science, but hopefully a case can be built one way or the other with a preponderance of evidence—to wit, is it, or was it, dead or alive?

Two other small devices that will be carried aboard the Mars 2020 rover deserve mention. One is a pair of small microphones, and they will be the first to record the sounds of Mars. The recording is unlikely to be dramatic—just the faint sounds of blowing wind and the occasional creaking of a rock under thermal stress, along with the crunching of the wheels as the rover drives across the surface—but this will be the first time we have heard the sounds of this world, and it will bring Mars one giant step closer to the minds of the public. It's a great PR move.

Mars 2020 will also carry a helicopter, another first for Mars. This small aircraft has been in development for years and has followed a rocky road of its own to gain approval—management has wondered if something that flies is really necessary? Given all the potential complications of such a device, will it work? Can it be built and tested in time to go along? In the end, the small drone was approved for inclusion. The main body of the Mars Helicopter is about the size of a cinder block, with two long contra-rotating helicopter blades sitting atop it. On top of these is a small solar panel that will facilitate charging its batteries after its short flights, during which it will provide imagery of the road ahead from up to a few hundred feet in altitude.

Fig. 34.1. JPL's Mars Helicopter, which, after years of struggle, made its way onto the Mars 2020 mission. The as-yet-unnamed Mars 2020 rover sits in the background, doubtless envious of the helicopter's ability to fly over the rugged surface. Image from NASA.

As for where the rover will be landed, this is still being debated. As in years past, the landing site selection process for the mission has been a complex push-and-pull process between scientists of differing stripes, engineers, and various NASA officials. A number of regions are still under consideration, and one must be picked before launch in 2020. The life science people want something likely to provide them with the best chance to find something alive—either in the past or today. The geologists want an area that has a rich diversity of rocks and geological processes to explore, as they had with Curiosity in Gale Crater. And the engineers, of course, want a place that is reasonably safe to land their machine. Make it as flat and safe (other disciplines would say "bland") as possible, please.

Toward this last goal, Mars 2020 has something helpful coming along for the ride called Terrain Relative Navigation, or TRN. In the past, Mars landers were committed to a landing zone once the probe was headed down toward the planet, based entirely on images acquired by NASA's Mars orbiters—no last-minute changes allowed, thank you very much. For Mars 2020, the TRN system will enable the rover to make last-minute decisions based on images taken during the seven-minute descent to the surface. The rover will compare those images to others in its computer memory that have been selected from NASA's vast archive of orbital photos. If the descending rover spots a disagreeable landform below—a large boulder, a crater rim, a sand dune, or some other mission-risky obstacle—the computer can divert the lander's trajectory and select a new landing site, subjecting it to the same level of scrutiny.

The rest of the landing sequence will proceed as did Curiosity's. At an altitude of about five miles, the landing radar will activate, giving the rover a running measurement of the range to the ground. Then, at one mile, the rover will separate from the top half of the aeroshell and at about 4,200 feet the sky crane's "backpack" rockets will fire to further decelerate the rover. Finally, at 65 feet above the surface, the spacecraft will slow to a hover, and will separate from the rocket pack to be lowered by its tethers. Once it is on the ground, a signal will be sent back up to the rocket pack, which will then separate to fly off and crash a few miles away. Mars 2020 will then be free to get situated and start exploring.

And where will it explore? The current leader in the small pack of candidate landing sites is Jezero Crater, or a region nearby. Jezero is an ancient crater that shows all the signs of having contained a lake in the past, much as Gale Crater did. But a nearby area, called Northeast Syrtis, has also been a prime candidate, due to its location between a large volcano and Jezero, and it is thought to be another area where, long ago, standing water could have been a breeding area for ancient life. Recently, a new region was added to the list of contenders, called Midway, referring to its location roughly halfway between Jezero and Northeast Syrtis, and some mission planners consider this to offer the best features of both regions. A final site decision will be made by 2019.

One more thing about Mars 2020—the wheels. As we have all seen, the wheels on Curiosity have been badly chewed up by the Martian terrain. They were designed to be light and strong, but the area between the wheels' cleats—called "grousers" when referring to Martian wheels—were too thin, and this almost foil-like part of the wheel was easily punctured and ripped. The wheels still work, thanks to a strong rim inside, but the design was not ideal. The new wheels are slightly smaller and are tougher—they have straight paddle-like grousers instead of the chevron-like ones on Curiosity, and thus fewer points of stress when deformed. They are made of thicker metal, are narrower and taller, and have been tested to withstand the very types of rock protrusions that damaged Curiosity's wheels—primarily sharp, vertical spikes left behind when softer Mars soil was eroded away and other sharply angled rocks.

Mars 2020's primary mission is expected to last two years, but with its robust, time-tested design and nuclear fuel supply, we can expect it to rove far longer. The discoveries are likely to be stunning, and if evidence of life is found, that will be a showstopper.

Fig. 34.2. The improved design for the Mars 2020 rover wheel. Stronger, narrower, and with fewer stress points when flexed, this wheel should be impervious to most of the damage suffered by Curiosity's wheels on Mars. Image from NASA.

Chapter 35

FLASH FORWARD: GRABBING THOSE SAMPLES

Leaving little sample tubes strewn across the surface of Mars is one thing, but getting them back to Earth is quite another. The first step in this process is to retrieve them so that they can be deposited into a return vehicle headed toward our home planet. It seems simple enough—just go and grab them with a robot—but it's harder than it sounds. That's why a project at JPL is working on determining every variable to assure success.

In a small lab, cluttered with bits and pieces of arcane-looking equipment, a team of robotics experts is testing every variation of a challenging sample-tube-grab on Mars that they can dream up. Retrieving the samples drilled by the Mars 2020 rover and left across the rugged terrain of the Martian surface will require great robotic dexterity and autonomous software, and we must understand how it can best be accomplished *now*, before the 2020 rover design is finalized and flown. Retrieving these samples is a task that will require careful grasping techniques, and both the sample containers and the future retrieval rover must be designed with this in mind—neither technology has yet been tested on distant worlds.

When the retrieval rover reaches Mars, many years will likely have passed since the Mars 2020 samples were cached. Some may be partially buried, some may have rolled up next to a rock or into a crevice, and some may have been deliberately placed under a rocky overhang to shield them from radiation exposure and temperature gradients.

Any of these conditions can challenge a simple tube-grab. If you've ever played one of those claw-grabber games in an arcade, trying to grasp a fat, slippery plush toy on a mound of other goodies with the smooth, metal claws, you know what I mean—it's always harder than it looks.

The Sample Transfer Technology team has considered a multitude of variables to enable a successful recovery in any scenario they can dream up. The most obvious element of the system is the "gripper," the robotic hand that will grasp sample tubes. This three-jawed robotic hand will grab the tube near one end. Feedback from the gripper will tell the onboard computer when just the right amount of force has been exerted to assure a secure grip on the sample tube—too much pressure might crush it, and too little could allow the tube to slip free.

This gripper will be at the end of a long robotic arm with six degrees of freedom. By comparison, Curiosity's arm had five degrees of freedom. The added degree of freedom—adding to the directions it can twist, turn, and rotate in—is required for the complex operations for sample-tube retrieval, as well as for loading the tube onto a Mars Ascent Vehicle for orbital rendezvous with the spacecraft destined to return it to Earth.

Accomplishing this task will require autonomous control software for the retrieval rover. It will need to be capable of automated obstacle avoidance while driving, detection of the sample tubes, and the collection and processing of the tubes. Since a single radio query from the rover can take as much as forty minutes to make the round trip to Earth and back, and tele-operation could extend operation time for a single sample-tube collection from a few hours to potentially several days, the more self-reliant the machine is the better.

Once Martian surface samples are available for analysis on Earth, we may learn more about the planet in a few months than we have during the past sixty years of exploration. A Mars sample-retrieval mission is the key to unlocking secrets about our most compelling planetary neighbor, and may give us answers to the question of life beyond Earth.

Chapter 36

OTHER MISSIONS OF EXPLORATION AND DISCOVERY

The next two decades of planetary exploration promise to be some of the most exciting ever. Not that the past sixty years have been dull—some of the most amazing accomplishments of the space age and beyond have been in robotic exploration. The first flybys of nearby planets, the first reconnaissance of the Martian surface, the frustrating but thrilling search for life on that planet, that first mad dash past the outer planets, and the large, complex machines sent there to survey Jupiter and Saturn in detail—these missions and others are truly amazing accomplishments. But, in a sense, as challenging as these missions were, each taxing the technologies of their time to the limit, they were the low-hanging fruit of the solar system. If any of these undertakings can be called easy, we've done them. The next phase of solar-system exploration involves reaching higher, aiming for ever more daring achievements, and doing so on smaller budgets, with more compact and technologically advanced machines. And the agencies who do this work, JPL premiere among them, have set the bar high—few would ever have imagined that a Mars rover could continue to be operational for fourteen years on the planet, but Opportunity continued to explore well into 2018 after a landing in 2004. We have come to expect much of our space machines.

Future plans for NASA and international space agencies could fill multiple volumes, so we will touch lightly on a select few here.

- Mars Sample Return: We've discussed JPL's plans for a sample-return mission to go and fetch the samples from the Mars 2020 rover and return them to Earth or possibly a space station. This will be a banner day for planetary scientists, answering questions about Mars we've had for at least two centuries. If life, current or past, is detected in the soil, that will be a huge bonus. It would challenge many long-held assumptions about life, the evolution of the solar system, and how humanity views itself. Fingers crossed that this mission gets approved and properly funded soon.[1]
- Exobiology on Mars (ExoMars): After launching a partially successful mission in 2016—the orbiter worked, but the lander failed—the European Space Agency is teed-up to fly another ExoMars mission in 2020. The new spacecraft includes a Russian-built lander and a European-sourced rover, which has goals generally similar to the Mars 2020 rover, without the sample caching component. The ExoMars drill is expected to extend as far as six feet into the Martian soil, and it will deliver these deep-core samples to an onboard laboratory for analysis. This lab has the ability to seek biosignatures and organic compounds in the soil, so this is a true astrobiology mission. The drill also carries an infrared spectrometer in the bit, so it can examine the bore-hole as it drills. Original plans included NASA in the effort, but the ever-hungry James Webb Space Telescope budget consumed the funds previously allocated for this, and NASA was forced to pull out of the partnership. Landing sites are still under discussion, and a final selection will be made in 2019.[2]

Europa Clipper: After more than a decade of consideration, NASA's Europa Clipper mission finally got the nod to move ahead. It's an orbiter set to launch in the early 2020s and will make at least forty-five orbital passes of Jupiter's most promising moon. Primary goals are to learn more about liquid water under the surface of Europa, the chemistry of that water and the icy surface, and the geological activity of the surface, along with surface interactions with the ocean below. It is not yet clear whether the orbiter's trajectory will allow it to fly through the plumes that blossom from Europa's surface, so planners are looking at the possibility of including a small fleet of CubeSats that might ride along and orbit the moon at low altitudes. Their data would be relayed to the Europa Clipper, recorded, and sent back to Earth. Some of these CubeSats may be optimized to seek biosignatures in the watery plumes, and could give us our first indication of life processes on an outer solar system world. A possible surface impactor has also been discussed, which would slam into the surface, releasing a shower of ice particles that could be collected by one of the CubeSats for analysis. Onboard instrumentation will include cameras, spectrometers (including one tuned to specifically seek organics on the surface), a subsurface penetrating radar, and a mass spectrometer to analyze Europa's tenuous atmosphere. Europa Clipper is a work in progress. Scheduling this mission depends partly upon the availability of NASA's Space Launch System rocket, the SLS, which is behind schedule—it's the only launcher deemed powerful enough to get the Europa Clipper to the Jovian system in the preferred timeline.[3]

Jupiter Icy Moons Explorer (JUICE): The European Space Agency is also designing a probe to explore Jupiter's moons. Set for launch in 2022, JUICE would not limit itself to Europa, but would also survey Callisto and Ganymede, both of which are also thought to have oceans below the surface. JUICE would end its mission in the 2030s in orbit around Ganymede. Russia is in discussions to partner on the mission with the Europeans, and their participation opens the possibility of using a Russian nuclear power supply instead of solar panels (which can be damaged by long exposure to Jupiter's radiation). The JUICE team is even looking at a possible lander add-on to descend to Ganymede and analyze the surface properties. Instrumentation includes cameras, spectrometers, and a surface-penetrating radar that would peer up to six miles below the ice surrounding these worlds.[4]

James Webb Space Telescope (JWST): While not a planetary probe per se, the JWST is a giant space telescope slated to launch in 2021. The sectional mirror will unfold after the JWST reaches its final position beyond the moon's orbit. JWST's mirror is 21.4 feet in diameter, and is truly a "big eye" when compared to the Hubble's eight-foot mirror; some truly amazing planetary science came from Hubble over the years, so the JWST should expand this reach dramatically. Goals include the observation of some of the oldest and farthest deep-space objects extant, beyond both our solar system and our galaxy, and the JWST will be capable of seeing back in time nearly to the origins of our universe. It will also observe exoplanets and provide the first direct imaging of some of the largest among them. The JWST is tuned to observe in long-wavelength visible light—orange to red—as well as in the infrared spectrum, allowing it to peer past much of the dust in intergalactic space that has obscured the view of visible wavelength instruments like the Hubble. The JWST will operate at very cold temperatures—necessary to observe in the infrared spectrum—shielded from the sun by a Mylar-like sunshield that makes it look like an Imperial Cruiser from *Star Wars*.[5]

Fig. 36.1. Artist's impression of the European Space Agency's ExoMars rover with the Russian-built lander in the background. Image from European Space Agency.

Double Asteroid Redirection Test (DART): The DART mission, while encompassing some scientific goals, is largely an Earth defense test. Our planet is under continual threat from objects in space, some large and some small, including asteroids and comets. While surveillance of these objects has increased substantially since 2010, when the Obama administration allocated a ten-times increase in funding to the effort, techniques to deflect them are still entirely theoretical. The DART mission will send a spacecraft to a Near-Earth Object (NEO) called Didymos, which is orbited by a tiny moonlet called, cutely, Didymoon. The primary goal of DART is to slam itself into the smaller Didymoon, which is about five hundred feet across, and see if the impact has any measureable effect on Didymoon's trajectory. If the experiment works, it would be the first step in the creation of a true Earth defense system, possibly allowing humanity to escape the fate of the dinosaurs. After all, we're supposed to be smarter than they were, so if we don't figure this out, we'll feel pretty foolish if an asteroid takes us out.[6]

PUFFER: With all this talk of large, expensive rovers headed toward Mars, a refreshingly novel approach has been undertaken by a team at JPL. Their mini-rovers are called PUFFERs, short for Pop-Up Flat Folding Explorer Robot. The little rovers are about the size of a large paperback book and would travel on the back of larger rovers traversing Mars. When the science team spots an area of interest that might be too risky to explore with a $2 billion rover, a PUFFER, which would be stored in a rack on the back of the rover, would be ejected. When it hits the ground, two wheels unfold from either end. The PUFFER has a small tail-dragger to keep it oriented while it drives around, investigating these riskier bits of terrain. PUFFER would carry a highly miniaturized instrument package with a camera and spectrometer at minimum. PUFFERs are fabricated from flexible, cloth-based circuit boards, and can be refolded and unfolded again via commands, so they are perfect for navigating into small crevasses, into craters, and even into lava tubes if found.[7]

- BepiColombo: This mission of the European Space Agency is

scheduled to arrive at the planet Mercury in 2025. It is a joint mission between ESA and the Japanese space agency, JAXA, and has a primary mission of one year. Two orbiters are planned. The first, the Mercury Planetary Orbiter (MPO) will provide the usual measurements and imagery of the planet, and the second, the Mercury Magnetospheric Orbiter (MMO) will map Mercury's magnetic field.[8]

• Luna-Glob: The Russians have a new lunar program planned, called Luna-Glob. The USSR had great success with their robotic exploration of the moon, including two robotic rovers in the 1970s. Luna-Glob has been under discussion since the mid-1990s, but due to budgetary constraints and political factors, it has been repeatedly delayed, with the latest estimate of an arrival at the moon between 2020 and 2025. It is unclear at this point exactly what the mission might encompass, but the Russian space agency is reported to be looking at a lander that would perform soil analysis and sample return from the lunar poles, an area of intense interest due to the possibility of water ice there.[9]

Chandrayaan-2: After orbiting the moon with its Chandrayaan-1, a lunar orbiter launched in 2008, India then sent an orbiter, Mangalyaan, to Mars in 2010. Both missions were successful on their first tries. India has plans to follow up each of them. Chandrayaan-2 is an orbiter/lander/rover combination that is planned for a lunar mission in 2018 or 2019, and the lander will deliver the rover at 70 degrees south latitude, the furthest south for any lunar landing probe to date. The rover will be capable of complex chemical analysis of rock samples found there. Mangalyaan-2, a follow-up to India's first successful Mars mission, is slated for a launch in 2020, and it may be just an orbiter or it may include a rover. The Indian Space Research Organization, or ISRO, is not specifying all their plans at the moment, but the country works fast and accomplishes such missions for far less cost than their western counterparts, in part due to lower labor costs. While previous missions were basically engineering projects, to prove the concepts behind robotic spaceflight, these new missions will be full research projects.[10]

Chang'e 4 and 5: China's lunar exploration program has progressed roughly in tandem with their new human spaceflight program. In 2003 China orbited Shenzhou 5, their first manned space capsule. By 2011, the country had orbited a small space station and visited it with Chinese *taikonauts*. In 2016, China flew a crew on their second space station—this is a country that, after decades of slow development, has begun moving quickly in spaceflight. In tandem with these human spaceflight missions China flew robotic projects to the moon. Chang'e 1, a lunar orbiter, reached the moon in 2007. Chang'e 3 landed a rover on the moon in 2013, which operated successfully—the rover kept working for three months, and the lander for a year. Chang'e 4 is a rover scheduled to be landed on the lunar farside—a first—in 2018 or 2019. Chang'e 5, a sample-return mission, is scheduled for launch in 2019. Its samples will be taken from almost seven feet below the surface. These are important missions for China's space agency, and they are designed as sequential steps toward human landings in the 2020s or 2030s.[11]

Fig. 36.2. China's Chang'e 3 rover, which operated on the lunar surface for three months. Image from China National Space Administration.

- The Small Lander for Investigating the Moon (SLIM): Japan is also headed to the moon . . . again. In 2007 the nation successfully sent their Kaguya (SELENE) orbiter to the moon, where it operated for a year and eight months, conducting research and sending back spectacular high-definition images. The main satellite was accompanied by two smaller satellites, proving Japan's ability to fly complex planetary missions with multiple spacecraft. This was the successor to the country's Hiten lunar orbiter, which flew in 1990. Japan also flew a comet rendezvous and sample mission called Hayabusa in 2003, that successfully returned a tiny sample of asteroidal material in 2010, after extensively studying the asteroid Itokawa. In 2021, JAXA plans to launch SLIM, a lander that will apply sophisticated "facial recognition" software to pinpoint a landing lunar site more accurately than anything since the Apollo landings in the 1960s and 1970s (which had human pilots at the controls). The planned landing site is the Marius Hills lava tube, a gaping hole in the moon that is thought to lead to a gigantic cavern. Japan is also involved with ESA's BepiColombo mission, and was a major contributor of hardware and cargo replenishment to the International Space Station.[12]

There are others—dozens of Earth orbital missions are planned, and others will visit comets and the sun, but the projects listed above are the major planetary and lunar programs. Asia is the newcomer, and the rising powers there are moving quickly.

The solar system seems to be getter smaller every year. While the promises of human bases on the moon and human missions to Mars, put forth decades ago by both the Soviet Union and the United States, are still just promises, the robotic explorations by these countries, and increasingly others, have made the solar system a place that in increasingly well understood. Only sixty years ago, the planets were places enshrouded in mystery, with most of our knowledge based on suppositions made from hazy telescopic observations. This has changed dramatically, with the pace accelerating in the past two decades as multiple international newcomers have joined exploratory efforts of the United States and Russia. We have now charted every planet we know of, including the former planet Pluto—now reduced to dwarf planet status with much acrimony (many people, astronomers among them, decried this change). We still have a long way to go—including that elusive first sample to be returned from another planet—but that achievement may not be far off.

Besides the armada of government spacecraft that have reconnoitered every corner of the solar system, private entrepreneurs are getting into the game. Elon Musk's company SpaceX has proven its ability to build and fly reusable rockets at a blistering pace, and Jeff Bezos, of Amazon fame, owns and operates Blue Origin, a company that is building a pair of rocket designs that will go into commercial operations in the next few years. Musk is currently building his Big Falcon Rocket, a massive machine—potentially more capable than even NASA's Saturn V moon rocket—with which he hopes to send people to the moon and Mars by the mid-2020s. Bezos is working on the robotic Blue Moon lunar lander that he plans to have in regular operation by 2022, delivering cargo to the moon and returning with whatever the space agencies—or other private companies—want to send back to Earth. It is an amazing time to be alive, as we watch a second space age dawn.

For the first time since spaceflight started in the 1950s, people from all walks of life are getting involved—space exploration is no longer the exclusive domain of large government agencies. CubeSats are being built by university students and garage innovators all over the world, and they will soon proliferate not just in Earth orbit but—as the new launch systems now in development become ever more affordable—around other planets as well. Other entrepreneurial operations are building robots to prospect and mine the moon and the asteroids. The space economy is set to move toward the $1 trillion range in the 2020s, and perhaps triple that by 2040, according to some studies.[13]

As Arthur C. Clarke wrote in his seminal science fiction novel *2010: Odyssey Two*, "All these worlds are yours . . ."[14] We might add, "use them wisely."

Chapter 37

THE CENTER OF THE UNIVERSE: PART 5

It's now September 15, 2017. I'm leaving JPL after watching the Cassini mission come to a quiet end . . . just the winking out of a tiny spike topping a green squiggly line plot on the big screen. We had all watched as the spike slowly diminished to nothing, leaving a flat green line. It was a small ending to a huge project, one that contributed greatly to our knowledge of Saturn, its moons, and the very origins of the solar system.

It's still dark when I leave the campus, and I am reminded of the night, five years earlier, when I left JPL after the successful landing of Curiosity. In a way these two warm summer nights are perfect bookends—one evening heralded a mission of planetary exploration that began in 2012, and another signaled the end of another incredible mission, one that revolutionized our understanding of Saturn and its moons, which ended in 2017. In just over a year from that time, I would be watching the dawning of another era, as I viewed Elon Musk's Falcon Heavy rise from Pad 39 at the Kennedy Space Center, to fly a near-perfect test mission. The Falcon Heavy ushers in a whole new ballgame; one which will lower the cost of launching large probes to deep space, enabling even more wonders to be discovered. What a decade it's been, and what amazing things are in store for the next one.

As I pass the guard kiosk and wave my badge, a red light goes on. I'm a bit dumbstruck, as I've never seen it light up before. A sleepy guard comes out of his shed—it's still early about 5:30 a.m., and the shift change has not yet arrived. He kindly informs me that he must take my badge. When I ask why, he shrugs and looks it up on the computer. It seems that it's been ninety days since my last contract with JPL expired, and all badges are repossessed at that time—whether you are a lowly media worker or a mission manager makes no difference, he tells me with a shrug. I turn it over with a sad smile and head off to my car. It looks like my era is over, too—for now. I'll be back to write about more technological wonders at the "Center of the Universe" soon enough, and like all those amazing spacecraft orbiting distant planets, transmitting their terabytes of data back to this wondrous place, I too will get mission extensions.

What a great time to be alive, I thought, and, as I slipped into the predawn glow, Mars shone, dim and red, on the horizon.

NOTES

CHAPTER 2: THE CENTER OF THE UNIVERSE: PART 1

1. Theodore Roosevelt, "The Strenuous Life," (speech before the Hamilton Club, Chicago, IL, April 10, 1899).

2. Mike Wall, "NASA's Jet Propulsion Laboratory (JPL): Facts & Information," Space.com, April 25, 2018, https://www.space.com/16952-nasa-jet-propulsion-laboratory.html.

CHAPTER 3: EARLY PLANS

1. US Army, *Project Horizon: Summary and Supporting Considerations*, vol. 1, June 9, 1959, US Army Center of Military History, https://history.army.mil/faq/horizon/Horizon_V1.pdf.

2. Mark Wade, "LUNEX," *Encyclopedia Astronautica*, http://www.astronautix.com/l/lunex.html.

3. "X-20 Dyna-Soar Space Vehicle," Boeing, https://www.boeing.com/history/products/x-20-dyna-soar.page.

4. Leonard Reiffel, *A Study of Lunar Research Flights*, vol. 1, Air Force Special Weapons Center, Kirtland Air Force Base, NM, June 19, 1959, http://www.dtic.mil/dtic/tr/fulltext/u2/425380.pdf.

5. Ibid., abstract.

6. Ibid., p. 2.

7. Ibid.

8. Ibid., p. 4.

9. Jeffrey T. Richelson, ed., "Soldiers, Spies and the Moon: Secret US and Soviet Plans from the 1950s and 1960s," National Security Archive, July 20, 2014, https://nsarchive2.gwu.edu/NSAEBB/NSAEBB479/.

10. "US Weighed A-Blast on Moon in 1950s," *Los Angeles Times*, May 18, 2000, http://articles.latimes.com/2000/may/18/news/mn-31395.

11. Matthew Gault, "Russia and America Had Plans to Attack the Moon with Nuclear Weapons," *National Interest*, July 31, 2017, https://nationalinterest.org/blog/the-buzz/russia-america-had-plans-attack-the-moon-nuclear-weapons-21734.

12. Joseph Myler, "Latest Red Rumor: They'll Bomb Moon," *Pittsburgh Press*, November 1, 1957, p. 25.

13. Robin Andrews, "This Is the Story of How America Once Thought About Nuking the Moon," *Forbes*, August 6, 2018, https://www.forbes.com/sites/robinandrews/2018/08/06/this-is-the-story-of-how-america-once-thought-of-nuking-the-moon/#4644193030b7; "US Considered Lunar A-Bomb Test," *Pittsburgh Post-Gazette*, May 18, 2000.

CHAPTER 4: *FRAU IM MOND*

1. Anatoly Zak, "German Legacy in the Soviet Rocketry," Russian SpaceWeb.com, http://www.russianspaceweb.com/rockets_german_legacy.html. Also Ernest G. Schwiebert, "USAF's Ballistic Missiles—1954–1964: A Concise History," *Air Force & Space Digest*, May 1964, p. 54.

2. Boris Chertok, *Rockets and People*, vol. 1 (Washington, DC: NASA Publications, 2005).

3. Letter from aerospace engineer and historian Matthew Polomik regarding impact speeds of early Soviet lunar probes, June 7, 2018; based on data from Don P. Mitchell, "Lunar Impact," Don P. Mitchell Homepage, 2008, http://mentallandscape.com/L_Luna2.htm.

4. Alan Bellows, "Faxes from the Far Side," Damn Interesting, October 23, 2015, https://www.damninteresting.com/faxes-from-the-far-side/.

5. Sydney Wesley Finer, "The Kidnaping of the Lunik," CIA Historical Review Program, September 18, 1995, Central Intelligence Agency Library, https://www.cia.gov/library/center-for-the-study-of-intelligence/kent-csi/vol11no1/html/v11i1a04p_0001.htm.

6. Ibid.

7. Ibid.; Amy Shira Teitel, "The CIA's Bold Kidnapping of a Soviet Spacecraft," *Popular Science*, October 20, 2015, https://www.popsci.com/cias-bold-kidnapping-soviet-spacecraft.

CHAPTER 5: SUICIDE ON THE MOON

1. John F. Kennedy, "Special Message to the Congress on Urgent National Needs," May 25, 1961, *American Presidency Project*, ed. Gerhard Peters and John T. Woolley, http://www.presidency.ucsb.edu/ws/?pid=8151.

2. John F. Kennedy, "Address at Rice University on the Space Effort," Houston, TX, September 12, 1962, http://explore.rice.edu/explore/kennedy_address.asp.

3. Edgar Cortright, oral history interview by Rich Dinkel, August 20, 1998, NASA/Johnson Space Center, JSC archives, https://www.jsc.nasa.gov/history/oral_histories/CortrightEM/CortrightEM_8-20-98.htm.

4. Ibid.

5. William H. Pickering, oral history interview by Mary Terrall, November 7–December 19, 1978, Caltech archives, http://oralhistories.library.caltech.edu/87/1/OH_Pickering_1.pdf.

6. James R. Hansen, *Spaceflight Revolution: NASA Langley Research Center from Sputnik to Apollo* (Washington, DC: National Aeronautics and Space Administration, 1995), https://history.nasa.gov/SP-4308.pdf.

7. John Johnson, "NASA's Early Lunar Images, in a New Light," *Los Angeles Times*, March 22, 2009.

8. Ibid.

CHAPTER 6: FLASH FORWARD: THE LUNAR FLASHLIGHT

1. Lee Mohon, ed., "Space Launch System," NASA, last updated October 9, 2018, https://www.nasa.gov/exploration/systems/sls/index.html.

2. A Falcon 9 launch currently costs about $62 million and can carry 50,000 pounds to orbit. This equates to about $1,240 per pound (given that most launches carry less than a full payload burden). A gallon of water weighs 8.34 lbs., so a gallon of water launched to orbit costs about $12,000. See SpaceX pricing tables: https://www.spacex.com/about/capabilities.

CHAPTER 7: MARS IN THE CROSSHAIRS

1. Claire L. Evans, "The Canals of Mars," *Universe*, ScienceBlogs, September 28, 2012, http://scienceblogs.com/universe/2012/09/28/the-canals-of-mars/.

To be more specific, early optics were uncoated (modern lenses are covered with a non-glare metallic coating), and were far more reflective than modern ones. Hence, the capillaries in the eye are sometimes reflected back onto the eyepiece lens when looking at a relatively bright object such as Mars.

2. Percival Lowell, *Mars and Its Canals* (New York: Macmillan, 1906), p. 376.

3. Robert Leighton, interview by the author, May 1995.

4. John Casani, interview by the author, April 2015.

5. Ibid.

6. A. J. Kliore, "Radio Occultation Exploration of Mars," in *Exploration of the Planetary System*, ed. A. Woszczyk and C. Iwaniszewska, International Astronomical Union, Symposium no. 65 (Copernicus Symposium IV), held in Torun, Poland, September 5–8, 1973 (Dordrecht: Springer Netherlands, 1974), pp. 295–316.

7. For more specifics on the planet Mars, see the NSSDC's "Mars Fact Sheet," by David R. Williams, last updated December 23, 2016, https://nssdc.gsfc.nasa.gov/planetary/factsheet/marsfact.html.

8. Robert Leighton, interview with the author, November 7, 1988.

CHAPTER 8: FLASH FORWARD: MarCO

1. "Mars Cube One (MarCO)," NASA/Jet Propulsion Laboratory, https://www.jpl.nasa.gov/cubesat/missions/marco.php.

2. Mark Betancourt, "CubeSats to the Moon (Mars and Saturn, Too)," *Air & Space*, September 2014, https://www.airspacemag.com/space/cubesats-moon-mars-and-saturn-too-180952389/.

CHAPTER 9: LANDING ON LUNA INCOGNITA

1. William H. Pickering, oral history interview by Mary Terrall, November 7–December 19, 1978, Caltech archives, http://oralhistories.library.caltech.edu/87/1/OH_Pickering_1.pdf.

2. This observation has been controversial, as the techniques used to examine the camera upon its return to Earth were not as sterile as they could have been. See "Surveyor 3 Streptococcus Mitis (APSTREPMIT)," NASA Life Sciences Data Archive, last updated January 17, 2018, https://lsda.jsc.nasa.gov/Experiment/exper/1651.

CHAPTER 10: RUSSIAN ROVERS

1. Richard Garriott, interview by Cindy Yans, "Lord British, We Hardly Knew Ye," *Computer Games Magazine*, April 13, 2001, https://web.archive.org/web/20050206145547/http://www.cdmag.com:80/articles/032/098/010413-i1.html.

CHAPTER 11: FLASH FORWARD: PROSPECTING THE MOON

1. The name of this lunar orbiting platform has changed a few times. First it was the "Deep Space Gateway," then the "Lunar Orbiting Platform." It was later rechristened the "Lunar Orbiting Platform-Gateway" or LOP-G, an inelegant appellation. NASA's recently appointed administrator, Jim Bridenstine, has more recently taken to simply calling it the "Gateway."

2. Google Lunar XPRIZE, 2018, https://lunar.xprize.org/.

3. Michael Wall, "Ex-Prize: Google's $30 Million Moon Race Ends with No Winner," Space.com, January 23, 2018, https://www.space.com/39467-google-lunar-xprize-moon-race-ends.html.

4. Doug Messier, "ispace to Launch Lunar Missions on SpaceX Falcon 9 in 2020, 2021." *Parabolic Arc* (blog), September 26, 2018, http://www.parabolicarc.com/2018/09/26/ispace-launch-lunar-missions-spacex-falcon-9-2020-2021/.

5. Jeff Foust, "Bezos Outlines Vision of Blue Origin's Lunar Future,"

SpaceNews, May 29, 2018, https://spacenews.com/bezos-outlines-vision-of-blue-origins-lunar-future/.

CHAPTER 12: THE CENTER OF THE UNIVERSE: PART 2

1. Rob Manning, interview with the author, June 2018.
2. Ibid.
3. Ibid.

CHAPTER 13: VOYAGERS ON MARS?

1. Some insight of this acrimony between NASA headquarters and JPL can be gleaned from a history of investigations into the failure of the Ranger lunar probes, R. Cargill Hall, "Chapter Eleven: In the Cold Light of Dawn," *Lunar Impact: A History of Project Ranger* (Washington, DC: Scientific and Technical Information Office, NASA History Series, 1977), https://history.nasa.gov/SP-4210/pages/Ch_11.htm.

2. Jay Gallentine, *Ambassadors from Earth: Pioneering Explorations with Unmanned Spacecraft* (Lincoln: University of Nebraska Press, 2014).

3. Much more detail can be found in the NASA historical publication, Edward Clinton Ezell and Linda Neuman Ezell, "Voyager: Perils of Advanced Planning, 1960–1967," in *On Mars: Exploration of the Red Planet, 1958–1978* (Washington, DC: Scientific and Technical Information Office, NASA History Series, 1984), https://history.nasa.gov/SP-4212/ch4.html.

4. David S. F. Portree, "The First Voyager (1967)," *Wired*, May 27, 2012, https://www.wired.com/2012/05/the-first-voyager-1967/.

5. Ezell and Ezell, "Voyager."

6. It's hard to say whether or not two Titan III launches ended up being much less expensive than a single Saturn V launch (as was planned for the final stages of Voyager Mars), as quoted numbers vary widely. But two separate launches did allow for one booster to fail and the other spacecraft to still have a successful conclusion.

CHAPTER 14: . . . AND THEN CAME VIKING

1. Edward Clinton Ezell and Linda Neuman Ezell, "Safe Havens: Selecting Landing Sites for Viking," in *On Mars: Exploration of the Red Planet, 1958–1978* (Washington, DC: Scientific and Technical Information Office, NASA History Series, 1984), https://history.nasa.gov/SP-4212/ch9.html.

2. Norman Horowitz, oral history interview by Rachel Prud'homme, Caltech Institute Archives, July 9–10, 1984, http://oralhistories.library.caltech.edu/22/1/OH_Horowitz_N.pdf.

3. Ibid.

4. Ibid.

5. Ibid.

6. Ibid.

7. Ibid.

8. For more stats on the Mariners, see NASA's NSSDCA's overview on the web: David R. Williams, "The Mariner Mars Missions," NASA, last updated January 6, 2005, https://nssdc.gsfc.nasa.gov/planetary/mars/mariner.html.

9. NASA Science Mars Exploration Program, "NASA Marks 30th Anniversary of Mars Viking Mission," news release, July 14, 2006, https://mars.nasa.gov/news/616/nasas-marks-30th-anniversary-of-mars-viking-mission.

10. Marisa Gertz, "This Is the First Photo Ever Taken from the Martian Surface," *Time*, July 18, 2016.

11. Ezell and Ezell, *On Mars: Exploration of the Red Planet*, https://history.nasa.gov/SP-4212/contents.html.

12. Ezell and Ezell, "On Mars," in *On Mars: Exploration of the Red Planet*, https://history.nasa.gov/SP-4212/ch11.html. Also, JPL, "Viking Status Report," 11:30 a.m. PDT, July 22, 1976; and Viking press briefing, July 21, 1976.

13. Ezell and Ezell, "On Mars."

14. Ibid.

15. Ibid.

16. Ibid.

17. Barry E. DiGregorio, "The Calm Before the Storm: An Interview with Dr. Gilbert Levin," *Mars Daily*, November 17, 2003, http://www.marsdaily

.com/reports/The_Calm_Before_The_Storm_An_Interview_With_Dr_Gilbert_Levin.html.

18. NASA Public Affairs Office, "NASA Confirms Evidence That Liquid Water Flows on Today's Mars," press release 15-195, September 28, 2015, https://www.nasa.gov/press-release/nasa-confirms-evidence-that-liquid-water-flows-on-today-s-mars.

19. Ibid.

CHAPTER 17: BARNSTORMING VENUS: PART 2

1. Rod Pyle, *Amazing Stories of the Space Age: True Tales of Nazis in Orbit, Soldiers on the Moon, Orphaned Martian Robots, and Other Fascinating Accounts from the Annals of Spaceflight* (Amherst, NY: Prometheus Books, 2017).

2. David F. Portree, *Humans to Mars: 50 Years of Mission Planning, 1950–2000*, NASA Monographs in Aerospace History, no. 21 (Washington, DC: NASA History Division, February 2001).

3. D. Cassidy, C. Davis, and M. Skeer, *Preliminary Considerations of Venus Exploration via Manned Flyby*, JPL History Collection, 15-437, Bellcomm Inc., November 30, 1967.

CHAPTER 18: FLASH FORWARD: THE TICK-TOCK ROVER

1. Andrew Good, "A Clockwork Rover for Venus," JPL News, August 25, 2017, https://www.jpl.nasa.gov/news/news.php?feature=6933.

2. Ibid.

3. Rob Latham, ed., *The Oxford Handbook of Science Fiction* (Oxford, UK: Oxford University Press, 2014), p. 441.

CHAPTER 20: A GRAND-ISH TOUR

1. Gary A. Flandro, "Fast Reconnaissance Missions to the Outer Solar System Utilizing Energy Derived from the Gravitational Field of Jupiter," *Astronautica Acta* 12, no. 4 (July 1966): 329–37.

2. Donald P. Hearth, interview by Craig B. Waff, August 7, 1988, cited in Craig B. Waff, "The Next Mission: Grand Tour or Jupiter-Intensive?" ch. 3 of *Jovian Odyssey: A History of NASA's Project Galileo*. unpublished manuscript, NASA Historical Reference Collection, pp. 6–7.

3. Andrew J. Butrica, "Voyager: The Grand Tour of Big Science," ch. 11 in *From Engineering to Big Science: The NACA and NASA Collier Trophy Research Project Winners*, ed. Pamela E. Mack, NASA History Series SP-4219 (Washington, DC: NASA History Office, 1998).

4. S. Ichtiaque Rasool, interview by Andrew Butrica, December 12, 1994, cited in Butrica, "Voyager."

5. John Casani, interview with the author, November 16, 2017.

CHAPTER 21: PIONEERING JUPITER AND SATURN

1. James Van Allen, "Pioneer Beat 'Warranty,'" *Aviation Week & Space Technology*, March 2003.

CHAPTER 22: JPL'S FINEST HOUR: THE VOYAGERS, PART 1

1. John Casani, interview with the author, November 16, 2017.
2. Ibid.
3. Ibid.
4. Ibid.
5. Ibid.
6. Ibid.
7. Edward C. Stone, interview by Andrew J. Butrica, November 23, 1994, cited in Andrew J. Butrica, "Voyager: The Grand Tour of Big Science," ch. 11 in *From Engineering to Big Science: The NACA and NASA Trophy Research Project Winners*, ed. Pamela E. Mack, NASA History Series SP-4219 (Washington, DC: NASA History Office, 1998).
8. Ibid.
9. Ibid.
10. Ed Stone and John Casani, "The Voyager Program's 40th Anniversary," lecture, American Institute of Aeronautics and Astronautics, November 16, 2017.

11. Casani, interview.

12. Ibid.

13. Stone and Casani, "Voyager Program's 40th Anniversary."

14. Ibid.

15. A. J. S. Rayl, "The Stories Behind the Voyager Mission: Linda Morabito Kelly," *Planetary Report*, September/ October 2002.

16. Overheard by the author at a Voyager press event in 1979.

17. Ian Morison, *A Journey through the Universe: The Gresham Lectures on Astronomy* (Cambridge, MA: Cambridge University Press, 2015), p. 74.

18. Stone and Casani, "Voyager Program's 40th Anniversary."

CHAPTER 23: BRIEF ENCOUNTERS: THE VOYAGERS, PART 2

1. Elizabeth Landau and Preston Dyches, "35 Years On, Voyager's Legacy Continues at Saturn," JPL News, August 24, 2016, https://voyager.jpl.nasa.gov/news/details.php?article_id=46.

2. Ed Stone and John Casani, "The Voyager Program's 40th Anniversary," lecture, American Institute of Aeronautics and Astronautics, November 16, 2017.

CHAPTER 24: INTO THE VOID: THE VOYAGERS, PART 3

1. Suzanne Dodd, interview with the author, August 22, 2013.

2. Ibid.

3. Ed Stone and John Casani, "The Voyager Program's 40th Anniversary," lecture, American Institute of Aeronautics and Astronautics, November 16, 2017.

4. To see a real-time mission clock and distance tracker, go to Jet Propulsion Laboratory's Voyager Mission Status page, https://voyager.jpl.nasa.gov/mission/status/.

5. Dodd, interview with the author.

6. Clarisse Loughrey, "Chuck Berry Dead: Read Carl Sagan's Letter to Berry on His 60th Birthday," *Independent*, March 19, 2017, https://www.independent.co.uk/arts-entertainment/music/news/chuck

-berry-dead-johnny-b-goode-carl-sagan-music-space-voyager-golden-records-a7637631.html.

7. "What Are the Contents of Golden Record," Jet Propulsion Laboratory: Voyager, https://voyager.jpl.nasa.gov/golden-record/whats-on-the-record.

CHAPTER 25: FLASH FORWARD: GOING INTERSTELLAR

1. Breakthrough Initiatives, https://breakthroughinitiatives.org.

CHAPTER 26: NOT BY A LONGSHOT

1. Paul Gilster, "Project Longshot: Fast Probe to Centauri," *Centauri Dreams* (blog), February 6, 2008, https://www.centauri-dreams.org/2008/02/06/project-longshot-fast-probe-to-centauri/.

CHAPTER 29: JUPITER'S REVENGE

1. Space Science Board, *Planetary Exploration, 1968–1975* (Washington, DC: National Academy of Sciences/National Research Council, 1968).

2. Emily Carney, "A Deathblow to the Death Star: The Rise and Fall of NASA's Shuttle-Centaur," *Ars Technica*, October 9, 2015, https://arstechnica.com/science/2015/10/dispatches-from-the-death-star-the-rise-and-fall-of-nasas-shuttle-centaur/.

3. "Frederick H. Hauck," interview by Jennifer Ross-Nazzal, NASA Johnson Space Center Oral History Project, November 20, 2003, https://www.jsc.nasa.gov/history/oral_histories/HauckFH/HauckFH_11-20-03.htm.

4. John Casani, interview with the author, November 16, 2017.

5. Karl Grossman, "The Risk of Cassini Probe Plutonium," *Christian Science Monitor*, October 10, 1997, https://www.csmonitor.com/1997/1010/101097.opin.opin.1.html.

6. Paul Weissman and Armand H. Delsemme, "Comet Shoemaker-Levy 9," Encyclopaedia Britannica, June 11, 2018, https://www.britannica.com/topic/Comet-Shoemaker-Levy-9.

7. Casani, interview with the author.

8. Torrence V. Johnson, "The Galileo Mission: Exploring the Jovian System," *Challenging the Paradigm: The Legacy of Galileo*," Proceedings of the 2009 Symposium and Public Lecture ((Pasadena: Keck Institute for Space Studies, California Institute of Technology, November 19, 2009), p. 25.

9. John Casani, interview with the author, April 30, 2016.

10. Ibid.

11. Johnson, "Galileo Mission," p. 26.

12. Ibid., p. 29.

13. Ibid., p. 30.

14. Ibid.

15. Jet Propulsion Laboratory "Mission News," October 16, 2009.

16. JPL press conference, September 21, 2003.

CHAPTER 30: FLASH FORWARD: INFLATABLE ANTENNAS

1. Rod Pyle, "JPL Technical Highlights, 2016–2017," Jet Propulsion Laboratory, California Institute of Technology, 2017, https://science andtechnology.jpl.nasa.gov/sites/default/files/documents/JPL_2016-2017 _Technology_Highlights.pdf.

CHAPTER 31: THE CENTER OF THE UNIVERSE: PART 4

1. Bobak Ferdowsi, interview with the author, March 27, 2014.

2. Bobak Ferdowsi, interview with the author, June 6, 2013.

3. Ibid.

4. Rob Manning, interview with the author, December 12, 2013.

CHAPTER 32: SATURN AHOY

1. To be completely fair, the Mars Observer mission was developed under the previous NASA administrator, Richard Truly, but Goldin's frugal methods are thought to be the primary reasons for the failure of

the MCO and MPL missions. For more detail on all three missions, see Rod Pyle, *Destination Mars: New Explorations of the Red Planet* (Amherst, NY: Prometheus Books, 2012).

2. Deborah Netburn, "'OK. Let's Do It!' An Oral History of How NASA's Cassini Mission to Saturn Came to Be," *Los Angeles Times*, September 12, 2017, http://www.latimes.com/science/la-sci-cassini-oral-history-20170912-htmlstory.html.

3. Ibid.

4. Ibid.

5. Ibid.

6. David Whitehouse, *Renaissance Genius: Galileo Galilei and His Legacy to Modern Science* (New York: Sterling Publishing, 2009), p. 100.

7. Ron Baalke, "Historical Background of Saturn's Rings," *Saturn Ring Plane Crossings of 1995–1996* (Pasadena, CA: Jet Propulsion Laboratory, 1999), https://web.archive.org/web/20090321071339/http://www2.jpl.nasa.gov/saturn/back.html.

8. Elizabeth Landau, "Fact Sheet: The Voyager Planetary Mission," Jet Propulsion Laboratory, https://voyager.jpl.nasa.gov/frequently-asked-questions/fact-sheet/.

9. David G. Simpson, "The Rings of Saturn," NASA Goddard Space Flight Center, Greenbelt, Maryland, https://caps.gsfc.nasa.gov/simpson/kingswood/rings/.

10. "Enceladus: Ocean Moon," Jet Propulsion Laboratory: Cassini Legacy, https://saturn.jpl.nasa.gov/science/enceladus (accessed May 2018).

11. David Mosher, "Saturn Ruled This Scientist's Life for 40 Years—Here's Why She Needs NASA to Go Back after Cassini's Death," *Business Insider*, September 17, 2017.

CHAPTER 33: FLASH FORWARD: DIVING ON TITAN

1. Michael Cole, "NASA Team Designing Sub to Explore Titan's Seas," *SpaceFlight Insider*, September 5, 2015, http://www.spaceflightinsider.com/space-flight-news/nasa-team-designing-sub-to-explore-titans-seas/.

CHAPTER 34: A RETURN TO MARS

1. Ken Farley and K. H. Williford, "Seeking Signs of Life and More: NASA's Mars 2020 Mission," *EOS*, January 11, 2017, https://eos.org/project-updates/seeking-signs-of-life-and-more-nasas-mars-2020-mission.

CHAPTER 36: OTHER MISSIONS OF EXPLORATION AND DISCOVERY

1. "Mars 2020 Rover," NASA, https://mars.nasa.gov/mars2020/.
2. "Robotic Exploration of Mars," ESA (European Space Agency), last updated May 2, 2016, http://exploration.esa.int/mars/48088-mission-overview/.
3. "Europa Clipper," Jet Propulsion Laboratory, https://www.jpl.nasa.gov/missions/europa-clipper/.
4. "JUICE," ESA, http://sci.esa.int/juice/.
5. Rob Garner, ed., "James Webb Space Telescope," NASA, last updated June 27, 2018, https://www.nasa.gov/mission_pages/webb/main/index.html.
6. Tricia Talbert, ed., "Double Asteroid Redirected Test (DART) Mission," Planetary Defense, NASA, last updated June 25, 2018, https://www.nasa.gov/planetarydefense/dart.
7. Andrew Good, "Origami-Inspired Robot Can Hitch a Ride with a Rover," Jet Propulsion Laboratory, March 20, 2017, https://www.jpl.nasa.gov/news/news.php?feature=6782.
8. "BepiColombo," ESA, http://sci.esa.int/bepicolombo/.
9. Anatoly Zak, "Luna-Glob Project," Russian Space Web, last updated December 11, 2017, http://www.russianspaceweb.com/luna_glob.html.
10. "GSLV-F10 / Chandrayaan-2 Mission," Department of Space, Indian Space Research Organisation, 2017, https://www.isro.gov.in/gslv-f10-chandrayaan-2-mission; Pallava Bagla, "India Eyes a Return to Mars and a First Run at Venus," *Science*, February 17, 2017, http://www.sciencemag.org/news/2017/02/india-eyes-return-mars-and-first-run-venus.
11. David R. Williams, "Future Chinese Lunar Missions," NASA Goddard Space Flight Center, last updated August 22, 2018, https://nssdc.gsfc.nasa.gov/planetary/lunar/cnsa_moon_future.html.

12. "Smart Lander for Investigating Moon (SLIM)," JAXA (Japan Aerospace Exploration Agency), http://global.jaxa.jp/projects/sat/slim/.

13. Michael Sheetz, "The Space Industry Will Be Worth Nearly $3 Trillion in 30 Years, Bank of America Predicts," CNBC News, October 31, 2017, https://www.cnbc.com/2017/10/31/the-space-industry-will-be-worth-nearly-3-trillion-in-30-years-bank-of-america-predicts.html.

14. Arthur C. Clarke, *2010: Odyssey Two* (New York: Del Rey, 1982), p. 320.

INDEX

Page numbers in **bold** indicate photos and illustrations.